Elias Hakalehto (Ed.)
Microbiology of Food Quality

Also of Interest

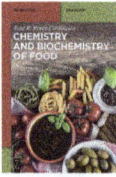

Chemistry and Biochemistry of Food
Jose R. Perez-Castineira, 2020
ISBN 978-3-11-059547-5, e-ISBN 978-3-11-059548-2

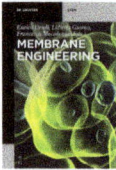

Membrane Engineering
Enrico Drioli, Lidietta Giorno, Francesca Macedonio (Eds.), 2019
ISBN 978-3-11-028140-8, e-ISBN 978-3-11-028139-2

Industrial Biotechnology.
Plant Systems, Resources and Products
Mukesh Yadav, Vikas Kumar, Nirmala Sehrawat (Eds.), 2019
ISBN 978-3-11-056330-6, e-ISBN 978-3-11-056333-7

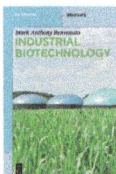

Industrial Biotechnology
Mark Anthony Benvenuto, 2019
ISBN 978-3-11-053639-3, e-ISBN 978-3-11-053662-1

Microbiology of Food Quality

Challenges in Food Production and Distribution During
and After the Pandemics

Edited by
Elias Hakalehto

DE GRUYTER

Editor
Adj. Prof. Elias Hakalehto
Finnoflag Oy
Isoharjantie 6
FIN-71800 SIILINJÄRVI
FINLAND
elias.hakalehto@gmail.com

ISBN 978-3-11-072492-9
e-ISBN (PDF) 978-3-11-072496-7
e-ISBN (EPUB) 978-3-11-072508-7

Library of Congress Control Number: 2021949679

Bibliographic information published by the Deutsche Nationalbibliothek
The Deutsche Nationalbibliothek lists this publication in the Deutsche Nationalbibliografie;
detailed bibliographic data are available on the Internet at http://dnb.dnb.de.

© 2022 Walter de Gruyter GmbH, Berlin/Boston
Cover image: m-gucci/iStock/Getty Images Plus
Typesetting: Integra Software Services Pvt. Ltd.
Printing and binding: CPI books GmbH, Leck

www.degruyter.com

Preface

Food is the most limiting factor for all humans. Sixty years ago, there was a strong debate about the possibility to feed the growing population. There were some 3–3.5 billion inhabitants on earth at that time, and 20% of these were living at a standard considered below "existence minimum" conditions. Since then, we have seen a dramatic increase in productivity concerning crop growth. United Nations and World Bank development indicators present figures on the development of many different aspects of life. One point of view refers to cereal yield measured as kg/ha in 1970 compared to the yield in 2009. The figures for low-income group are 1,296 and 1,952 respectively. For middle-income groups, the corresponding figures are 1,515 and 3,202 and for high-income countries 2,766 and 5,448 kg/ha, Thus, we see a doubling during these 40 years not only for middle- and high-income countries, but also for low-income countries, there is a significant increase in yield. This has been resulting from the increased use of fertilizers, irrigation, and more productive species of different crops.

To this, we have seen a strong decrease in food waste as cereals are kept away from rats and similar in a better way, but the logistic system has been developed dramatically as well! This preserves food and other food in a much better way. Smashed meat, for instance, now can be used after a week, while it had to be eaten within a couple of days earlier. In addition, we have much more dried and frozen food, and freezers are now common in most countries.

What we now see developing is a new type of food like fermented food where microorganisms are used to preserve food. Still, microorganisms can also degrade cellulosic material, which has not been possible except when converted in the cows' digestion system. Also, here microorganisms are used, but they are also causing problems with greenhouse gases like methane. Now we can see possibilities to avoid the negative effects of methane. One alternative is to absorb methane in, for example, activated carbon, from where we can strip it off and use it – for example, electricity and heat production, or as vehicle fuel. Another interesting method may be to feed the cows with red algae, which has proven to reduce methane formation in experiments in Australia. Also, bio-digestion of waste in different ways has been developing a lot last 20 years and hopefully will proceed further. We also see how larvae can convert food waste into protein-rich food or fodder. An example of this is the 1 ton/d plant in Eskilstuna, where soldier fly larvae convert food waste with a conversion efficiency of some 30–35%. At the same time, the rest material is used for biogas production and used as fertilizer in farmland. This is showing how we can get a circular society. First, we reduce the waste as such by better handling and logistics, including storage. Then we utilize the waste, and we still get to produce valuable products. This is system optimization!

Another important aspect has come to our minds recently; that is, pandemics and how to address this in the future. Large storages were made for some safety

https://doi.org/10.1515/9783110724967-202

materials in Finland, but still these were gone within few weeks. In other countries like Sweden, all storage was destroyed 20 years ago as "they would not be needed in the future." How shall we prepare society for a similar situation as with COVID-19 in the future? What storages are needed – food, safety equipment, fuel, etc.? How can we avoid spreading viruses and bacterial diseases via food or by their deliveries? Another aspect is how we can utilize microbes to upgrade side streams from food industries and agriculture. Biorefineries can be used to produce basic chemicals that can be further applied for producing all materials we need.

We still have just shown the possibilities, but shortly, we hopefully will see many of these possibilities also implemented in a large scale. Then we will both reduce the negative impact on climate and economy and see that we optimally utilize our limited resources. There is a trade-off between maximal production of crops and keeping the biodiversity at a high level. The distribution of fertilizers can be optimized and waste recirculated. New species with higher resistance to draft and biocontrol insects and other diseases will reduce the need for biocides. We still should have the possibility to produce what food we need. There will also be a balance between eating crops directly or taking the route over animals. It is not all meat or no meat, but the proportion between meat and crops for sustainability. Good hygiene, especially in the logistics around animal breeding, is a necessity that includes minimizing antibiotics to avoid antibiotic-resistant microorganisms. This demands careful care of our domestic animals. Cows, sheep, pigs, and hens should be treated principally as we treat our cats and dogs! There are many good examples of this, and let us see that regulations are driving in a positive direction for all!

With this I would very much encourage the readers to use the book and think about how you can impact the future from your own perspectives. Happy reading!

Erik Dahlquist
Senior professor in Energy Technology
School of Business, society and engineering
Malardalen University, Vasteras, Sweden

Contents

List of contributing authors

Hanna-Leena Alakomi
Technical Research Centre of Finland VTT Ltd
Espoo
Finland

Satu Salo
Technical Research Centre of Finland VTT Ltd
Espoo
Finland

Elias Hakalehto
Finnoflag Oy, Kuopio and Siilinjärvi, Finland;
Department of Agricultural Sciences,
University of Helsinki, Helsinki, Finland;
University of Eastern Finland, Kuopio, Finland

Mikko Immonen
Department of Food and Nutrition,
University of Helsinki, Finland

Ndegwa H. Maina
Department of Food and Nutrition,
University of Helsinki, Finland

Rossana Coda
Department of Food and Nutrition,
University of Helsinki, Finland

Kati Katina
Department of Food and Nutrition,
University of Helsinki, Finland

Frank Adusei-Mensah
Finnoflag Oy, Kuopio and Siilinjärvi, Finland
Department of Public Health,
Institute of Public Health and Clinical
Nutrition, University of Eastern Finland,
Kuopio, Finland.

Carina Tikkanen-Kaukanen
Ruralia Institute and Helsinki Institute of
Sustainability Science, University of Helsinki
Finnish Organic Research Institute,
University of Helsinki
Mikkeli, Finland

Anneli Heitto
Finnoflag Oy
Kuopio and Siilinjärvi, Finland

Ari Jääskeläinen
Savonia University of Applied Sciences
Kuopio, Finland

Jukka Kivelä
Department of Agricultural Sciences,
University of Helsinki
Helsinki, Finland

Jan den Boer
Department of Applied Bioeconomy,
Wrocław University of Environmental and
Life Sciences, Poland

Emilia den Boer
Faculty of Environmental Engineering,
Wrocław University of Science and
Technology, Poland

Alli Pesola
Finnoflag Oy, Kuopio and Siilinjärvi, Finland
Institute of Biomedicine, University
of Eastern Finland, Kuopio, Finland

Jouni Pesola
Department of Paediatrics, Kuopio University
Hospital, Kuopio, Finland;
School of Medicine, University of Eastern
Finland, Kuopio, Finland

Robert Armon
Division of Environmental, Water and
Agricultural Engineering,
Technion
Israel Institute of Technology, Haifa, Israel

Jukka M. Sauramäki
Posti Oy, Helsinki, Finland

Jukka-Pekka Hakalehto
Finnoflag Oy, Kuopio and Siilinjärvi, Finland;
Institute of Dentistry, University of Eastern
Finland, Kuopio, Finland

https://doi.org/10.1515/9783110724967-204

Elias Hakalehto

Introduction

Food is much more than a mixture of nutrients. It contains substances and microbes, which react with our body system and its microbiome. All these interactions influence the nutrition of our cells and tissues. It gives them the necessary energy and molecular building blocks for maintaining life.

It is a real danger in everyday life that our food materials get contaminated or transmit undesired chemical components or microbes. These contaminants can compromise our health and well-being, and their eradication requires proper manufacturing, storage, transportation, and preparation of various foods. It also warrants rules and regulations, as well as their thoughtful implementation.

Industrial monitoring of food quality, preservation methods, and other means of food maintenance are actually methods for avoiding spoilage and contamination (Hakalehto, 2016). The accomplished safety is a reward of active measures. They are based on scientific research and practical experience within various branches of industries. Although each food material has its own characteristics and typical microflora, the principles of hygiene maintenance are rather similar in the entire food and nutritional or environmental health sector.

In hygiene monitoring, it is just not enough to detect harmful or hazardous organisms. Still, it is important to understand their functions in the unseen world (not directly observable by our senses). In fact, as researchers and other professionals are dealing with microorganisms, they should conceive themselves at the microscopic scale. For example, the small size of the bacterial cells or viral particles allows their proportionally vast surface areas. Consequently, the microbes have lots of surfaces to attach, agglutinate, metabolize, transfer signal molecules or genetic materials, take up nutrients, enter the host cells, and so on.

To fully comprehend their activities, we need to understand the effects of scale. Correspondingly, our sensory views on reality cannot always reflect the functions of the microbial world. However, we need the "downgraded vision" for realizing the physical movements and forces influencing the microbial particles. For instance, a particle with a diameter less than 10 μm is not landed by gravity but randomly moves in the airspace along the fluctuations and rainfalls. Due to both the high importance of microbial security and the deviant scale of microbial world we need constant development of techniques for monitoring and controlling the food macroscopic and microscopic composition. Rapid and reliable microbial detection is of crucial importance.

Elias Hakalehto, Finnoflag Oy, Kuopio and Siilinjärvi, Finland; Department of Agricultural
Sciences, University of Helsinki, Helsinki, Finland; University of Eastern Finland, Kuopio, Finland

https://doi.org/10.1515/9783110724967-001

Besides the spatial organization of microbial cells, particles, communities, aggregates, mycelia, spores, or biofilms, it is essential to understand their physiological status and functions. These determine and illustrate the basic principles of biology as it is eloquently demonstrated in the book *The Microbe's Contribution to Biology* by Dutch microbiologists A.J. Kluyver and C.B. van Niel (1956). The book is based on the lectures given by the authors at Harvard University in April 1954. In their survey, the metabolic boundaries for microbiological activities have been fundamentally probed, and the microbial kingdom is seen as it is: as an essential dimension of the global ecosystems and as a model for metabolic activities. These metabolic potentials we have also seen in a very concrete way in the numerous bioprocessing projects conducted by my company, Finnoflag Oy. In fact, the biorefinery enterprises could broaden the concept of food and extend our possibilities of producing it globally.

During my studies in the Bioengineering Department of the University College London in 1984–85, it became most evident that the prevention of microbial contamination is a highly demanding task. During the lectures of Professors Malcolm D. Lilly FRS., and Peter Dunhill FRS., we also learned that "downstream processing is a losing game." In other words, we are prone to get defeated in our efforts to both control the spread of microbes around us and in purifying their metabolic products unless we put our unlimited focus on the combination of microbial technology with the common mainstream industries and services. The produce thereof includes the products of the biorefineries, which in the future will be decisive for our nutrition, as the population is continuously increasing. Biotechnology also offers means for healthiness for the global population.

Since the beneficial microbes are crucial for our food production, we need to understand the ways how to protect our interaction with them in a good sense, as well as to safeguard the food distribution against the harmful contamination. Fortunately, we can rely on the work of earlier generations of scientists. The past chairman of the Scientific Committee of the Finnish Academy of Sciences, and Professor of Microbiology of the University of Helsinki, late Helge Gyllenberg stated in his book (1958) that "food microbiology is largely microbial ecology applied to foodstuffs."

In the current situation with global pandemics, we have lived exceptional times by many means and measures. For example, it is now a foreseeable future that automobiles may be replaced by flying vehicles (Hawkins, 2017). Various drones are already in service for the distribution of food portions. Bringing food with them is definitely depriving the direct human touch from the catering services. We have to develop the deliveries in the air, land, and sea. This should be our responsibility as food sector professionals.

At the same time as we have huge concerns about the global development, for example, in areas like Africa, there is a glimpse of hope, as the novel technologies and microbiological knowledge could be harnessed to protect the common good. For example, at the same time, we need to make sure that new epidemics will not be provoked if possible, for example, by humans carelessly intruding to the caves of the bats for mining purposes or fervently destroying their forests. If the ecological principles are not respected, harmful viruses and microbes will rampage this planet and that is most enervating for Mankind.

The author wrote in his blog on the LedFuture Oy's homepage on February 18, 2021 as follows:

> In the current worrying global health situation, it would be most advisable to learn from history; to enhance the research for finding new antibiotics, antimicrobial substances, production of passive immunization means as suggested by Hakalehto and Kuronen some 25 years ago when proposing the use of chicken egg yolk antibodies against the virus epidemics, (Hakalehto and Kuronen, 1998) finding of novel probiotics as the continuation of *Lactobacillus rhamnosus* GG strain which was studied and launched by Valio Corporation, Finland, laboratories in Helsinki. Then we hit a vein of 'the gold of health', which could be further researched and extended by finding novel microbial strains and communities. By the way, it has been proven that microbiome balance could help to also resist SARS-CoV-2 viruses and many other harmful germs.
>
> (Hakalehto, 2021)

We should also implement all the developing new technologies to assist hygiene control in the field of food services. The fairer and healthier production and distribution of food should be the main objective.

References

Gyllenberg, H. G. (1958). Elintarvikkeiden pieneliöt: Niiden toiminta, torjuminen ja hyväksikäyttö. (The micro-organisms in foods: Their activities, prevention and utilization) (In Finnish). Pellervo-seura.

Hakalehto, E. (2016). Microbiological surveillance methods for the industries: comments on general strategies and theoretical background. In: Hakalehto, E. (ed.). Microbiological Industrial Hygiene. New York, NY, USA: Nova Science Publishers, Inc., 2016, 209–230.

Hakalehto, E. (2021). Past strive for safety innovations paving the way to sustainable health future. In: UVC-LED Blog & News. Led Future Oy. First published on February 18, 2021.

Hakalehto, E., Kuronen, I. (1998). A method for producing jelly sweets, which contain antibodies. WIPO Patent Application WO/1998/043610.

Hawkins, A.J. (2017). Watch this all-electric 'flying car' take its first test flight in Germany. https://www.theverge.com/2917/4/20/15369850/lilium-jet-flying-car-first-flight-vtol-aviation-munich.

Kluyver, A. J. and van Niel, C. B. (2013). The Microbe's Contribution to Biology. Cambridge, MA and London, England: Harvard University Press. Original publication date: January 1, 1956. https://doi.org/10.4159/harvard.9780674188693

Hanna-Leena Alakomi, Satu Salo

1 Microbiological quality and safety – a general overview

Abstract: Monitoring and safeguarding microbiological quality and safety of drinks and foods is essential for health. Good manufacturing practices should be practiced during the whole production process, from raw materials to the end product, as well as in every phase between them. Microbial contamination can readily take place on every occasion where the process is disturbed for some reason. Efficient monitoring of the quality of the foodstuff at different steps of the process is important to avoid foodborne infections and even epidemics that may cause morbidity and mortality in large numbers of consumers within a short period of time.

1.1 Introduction to microbiological quality

Microbiological quality and safety of food is a sum of several factors. Good quality of raw materials, process hygiene, including both clean air and surfaces, and personnel having good manufacturing practices are key factors in the production of safe food. Microbiological spoilage has been defined as a process or change, which renders a product undesirable or unacceptable for consumption (Nychas and Pangou, 2011). In addition, ability of microbes to grow and spoil the food is influenced by many factors, including chemical composition of the product, free nutrients, microbial interactions, microbial enzymatic activities, water activity, pH, and storage conditions like moisture and temperature as well as storage time.

Contamination of food by microbial agents is a global public health concern (WHO, 2013). Access to safe and sufficient food is a basic human necessity. Food spoilers can cause great economical losses, and zoonotic microbes can cause foodborne diseases. According to the World Health Organization (WHO), an estimate of 600 million (almost 10% of people in the world) people fall ill after eating contaminated food, that is, 420,000 people die annually, resulting in the loss of 33 million healthy life years (https://www.who.int/news-room/fact-sheets/detail/food-safety). In addition, US $110 billion is lost annually in productivity and medical expenses resulting from unsafe food in low- and middle-income countries. Unsafe food affects particularly infants, young children, elderly, and people with immunocompromised immune systems. Diarrheal diseases cause 550 million people to fall ill and 23,000 deaths annually.

Hanna-Leena Alakomi, Satu Salo, Technical Research Centre of Finland VTT Ltd, Espoo, Finland

https://doi.org/10.1515/9783110724967-002

According to the One Health approach, the health of people is connected to the health of animals and our shared environment (https://www.cdc.gov/onehealth/ba sics/index.html). Therefore, it is important to promote collaboration across all sectors. One Health approach includes zoonotic diseases, antimicrobial resistance, food safety and security, vector-borne diseases, environmental contamination, animals, and the environment.

According to the European Food Safety Authority (EFSA) and European Centre for Disease Prevention and Control (ECDC) report (2021), campylobacteriosis was the most reported gastrointestinal infection in humans in the European Union (EU) and has been so since 2005. Salmonellosis is the second most common reported gastrointestinal infection in humans and an important cause of foodborne outbreaks in the EU. Table 1.1 summarizes examples of foodborne pathogens. Foodborne cases of *Campylobacter* are mainly caused by raw and undercooked poultry, raw milk, and contaminated drinking water.

Table 1.1: Examples of health hazards associated with harmful microbes.

Microbes	Hazards	Influence of humans	Examples of species
Molds	Mycotoxins	Severe chronic and acute toxicity	*Aspergillus, Penicillium*
Bacteria	Foodborne infections and intoxications, allergic reactions	Diarrhea, vomiting, fever, others like pathogenic *Listeria sp.* – miscarriage in pregnant women	*Escherichia coli* *Salmonella sp.* *Listeria monocytogenes* *Staphylococcus aureus* *Vibrio cholerae*
Sporeforming bacteria			*Bacillus cereus* *Clostridium botulinum*
Viruses	Waterborne infections	Gastroenteritis, liver inflammation	Hepatitis A, norovirus, rotavirus
Protozoa	Human infections	Gastrointestinal illness	*Cryptosporidium parvum, Cyclospora sp.*

In nature and in process environments, microbes grow as multispecies communities (biofilms) attached to surfaces where they are protected from the action of biocides and disinfectant agents (Flemming et al., 2016). Cells of the community have close interactions with each other (multiple attachments and cell–cell signaling) and they cooperate for obtaining nutrients and metabolic compounds. In addition, cells in biofilm exhibit an altered phenotype (e.g., slower growth rate and gene expression) compared to free-swimming (planktonic) cells, and biofilm structure provides the cells with shelter from an exposure to external agents (e.g., biocides, disinfectant agents, and antibiotics; Flemming et al., 2016). In particular, extracellular polysaccharides play various roles in the formation of structures and the function of different biofilm communities: they exclude and/or influence the penetration of antimicrobial agents and provide protection against a variety of environmental stresses such as UV radiation, pH shifts, osmotic shock, and desiccation (Rumbaugh and Sauer, 2020).

1.2 Importance of process hygiene

The central goal of food safety policy is to ensure a high level of protection of human health regarding the food industry. The safety of the food is supervised by the health authorities. In Europe, the European Commission aims to assure a high level of food safety within the EU through coherent Farm to Fork measures and adequate monitoring. The website "ec.europa.eu/food" https://ec.europa.eu/food/index_en managed by the Directorate General for Health and Food Safety is describing the implementation of integrated Food Safety policy in the EU. It involves following actions: (a) to assure effective control systems and evaluate compliance with EU standards in the food safety and quality, animal health, animal welfare, animal nutrition, and plant health sectors within the EU and in non-EU countries in relation to their exports to the EU; (b) to manage international relations with non-EU countries and international organizations concerning food safety, animal health, animal welfare, animal nutrition, and plant health; and (c) to manage relations with the EFSA and ensure science-based risk management.

Scientists worldwide are reporting emerging pathogens found in foods or food production facilities (Wirtanen and Salo, 2005; Hammond et al., 2015). This shows that the fight against microbes is highly important to assure food safety. Despite this, an overall detection protocol for tracing microbes in food production chain is still not perfect. Most of the detection methods used in laboratory studies have been successfully applied to pilot scale and industrial field studies though evidently there are several limiting factors in the methods used. For instance, taking representative hygiene sample from inner surface of pipeline using cotton-tipped swab is inefficient. The quality control personnel should choose the proper means to remove the microbes from process environment by using scientific-based knowledge. Different microbe populations adapt differently to various food processing environments, and therefore case studies are needed to obtain detailed knowledge on microbe removal through optimized cleaning strategies.

Likewise, the plant layout and the production procedure play an important role in the hygienic state of the factory. As an example, one of our studies showed that returned milk transportation crates functioned as a vehicle transporting *Listeria sp.* into high hygiene areas of a dairy. Rearrangement of some process steps and improvement of the cleaning procedure based on ultrasonication were found to improve the hygiene in the case study (Salo and Wirtanen, 2007). In this context, also the combination of equipment design and the cleaning operation is of interest. For example, large fermentation tanks need different cleaning and sampling strategies compared to the milk transportation crates.

1.3 Process and equipment design in food industry

Poorly designed equipment, which are difficult to clean, can harbor spoilage microbes and pathogens. This can be avoided through hygienic equipment design, which should be taken into consideration when purchasing new or redesigning old process lines. The basic rule in hygienic design is to use simple structure, which is stated in most common guidelines. Detailed guidelines for hygienic design are available, for instance, from European Hygienic Engineering and Design group (EHEDG) (https://www.ehedg.org/guidelines/). EHEDG guideline doc 8 defines hygienic design principles. A hygienic design of process equipment has a tremendous impact on diminishing the risks of contamination of foods during production and hence on the products' shelf life. If the process equipment is of poor hygienic design, it is difficult to clean it from microbes. The microbes may furthermore survive and multiply in the crevices and dead areas of the equipment or process line. Poor hygienic design of process equipment and components used in the food processing industry is a risk for food contamination. The hygienic design of process equipment and components should be based on a sound combination of process and mechanical engineering as well as knowledge in microbiology. With a good hygienic design, the lifetime of the equipment will increase, while the maintenance and the manufacturing costs will be reduced.

The safety of food products manufactured with equipment meeting EU legislation and regulations has improved. The choice of materials and their surface treatments, for example, grinding and polishing, are important factors in inhibiting the formation of biofilm and in promoting the cleanability of surface. The process equipment is easy to clean if the surface materials are smooth and in good condition. Dead ends, corners, cracks, crevices, gaskets, valves, and joints are vulnerable points for biofilm accumulation.

1.4 Air handling systems

The quality of air in food production facilities is very important for the final product quality. The microbial population in the air channels depends on the environment, filtration membranes, and the sites of air holes. Formation of biofilm in air-conditioning systems does not occur without a water reservoir of some kind. Normally, there is no water in the air-conditioning systems, but it can accumulate unintentionally through condensation. When the air-conditioning system is to be cleaned and disinfected, it is very important that the disinfection medium penetrates the biofilm and does not simply flow through the system with the air.

1.5 Hygiene assessment in food processing environment

Harmful microbes may enter the manufacturing process and reach the end product in several ways, for example, through raw materials, air in the manufacturing area, cleaning tools and chemicals, process surfaces, process water, or factory personnel. Monitoring of microbes in raw materials, on surfaces, in the air and in final products is important in food processing. The cleanliness of food contact surfaces is the first goal, but it is important to keep the level of microbes as low as possible in the whole food processing environment since cross contamination can appear. Assessment of process hygiene can be based on observation of technical solutions from a hygienic design point of view, on soil accumulation in processing lines, or on detecting microbes. The interpretation of results obtained in liquid flow and on surfaces needs to be based on holistic risk assessment since official recommendations are scarce. Failures in cleaning can be noticed by comparing obtained individual results to results from long-term routine hygiene control.

1.6 Conclusions

It is important that consumers can rely on the safety of their beverages and foods. Awareness of the microbial presence and the risk of contamination at every step of the food manufacturing processes as well as improving the process hygiene and continuous monitoring are key for maintaining high standards of food quality.

References

EFSA and ECDC (European Food Safety Authority and European Centre for Disease Prevention and Control). (2021). The European Union One Health 2019 zoonoses report. EFSA Journal, 19 (2): 6406, 286. https://doi.org/10.2903/j.efsa.2021.6406.
Flemming, H.-C., Wingender, J., Szewzyk, U., Steinberg, P., Rice, S.A., Kjelleberg, S. (2016). Biofilms: An emergent form of bacterial life. Nature Reviews Microbiology, 14: 563–575. https://doi.org/10.1038/nrmicro.2016.94.

Hammond, S.T., Brown, J.H., Burger, J.R., Flanagan, T.P., Fristoe, T.S., Mercado-Silva, N., Nekola, J.C., Okie, J.G. (2015). Food spoilage, storage, and transport: Implications for a sustainable future. BioScience, 65 (8): 758–768. https://doi.org/10.1093/biosci/biv081.

Nychas, G.-J.E., Panagou, E. (2011). Microbiological spoilage of foods and beverages. In: Kilcast, D., Subramaniam, P. (eds.). Food and Beverage Stability and Shelf Life. Woodhead Publishing Series in Food Science, Technology and Nutrition. Cambridge, UK: Woodhead Publishing, 3–28. ISBN 9781845697013, https://doi.org/10.1533/9780857092540.1.3

Rumbaugh, K.P., Sauer, K. (2020). Biofilm dispersion. Nature Reviews Microbiology, 18, 571–586. https://doi.org/10.1038/s41579-020-0385-0

Salo, S., Wirtanen, G. (2007). Ultrasonic cleaning applications in dairies: Case studies on cheese moulds and milk transportation crates British Food Journal, 109 (1): 31–42.

WHO. (2013). Advancing food safety initiatives: Strategic plan for food safety including foodborne zoonoses 2013-2022. 1.Food safety. 2.Food contamination. 3.Foodborne diseases. 4.Health planning. I. World Health Organization. ISBN 978 92 4 150628 1 (NLM classification: WA 701) https://www.who.int/publications/i/item/9789241506281.

Wirtanen, G., Salo, S. (2005). Biofilm risks. In: Lelieveld, H.L.M., Mostert, M.A., Holah, J. (eds.). Handbook of Hygiene Control in the Food Industry. Cambridge, UK: Woodhead Publishing Ltd., 46–68.

Hanna-Leena Alakomi, Satu Salo, Elias Hakalehto

2 Preservation techniques, storage, stability, traceability, and other means to maintain food quality during the distribution chain

Abstract: The chain from the culturing, production, packaging, storage, and transportation of food includes numerous steps from original source to the tables of the consumer. These food services, to be safe and successful, demand a lot of expertise and professionalism. Most people live in cities, where food comes in a collective and centralized manner. Therefore, it is crucially important to follow the chain and monitor food quality all the way. This chapter describes the basic concepts of tracking the food and maintaining its quality. Proper preservation techniques are one of the keys for successful food deliveries. Many items require cold chain, or otherwise specific conditions with respect to humidity, light, or other parameters. Correct keeping, handling, and distribution of the foodstuff are well regulated during normal times. In the pandemic era, extra preparation, caution, and control are most necessary additional measures.

2.1 Preservation techniques: an introduction

Aim of food preservation is to prevent and reduce food loss in the production system as well as extend product shelf life. An efficient preservation prevents both autolysis (e.g., enzymatic activities) and microbial growth.

Food industry is a traditional industrial sector with several small and medium size enterprises. Implementation of novel food processing technologies is regulated by the national and international authorities. Hence, application of new processing technologies is often a slow process. Food safety and consumer acceptance play an important role in the selection of new food processing techniques.

Nonthermal food processing treatments like pulsed electric field, ultra-high pressure, high electric field pulses, and light pulses have gained a lot of attention during recent years. The demand by consumers for high-quality foods having "fresh" or "natural" characteristics has led to the development of foods that are preserved using mild technologies (Table 2.1).

Hanna-Leena Alakomi, Satu Salo, Technical Research Centre of Finland VTT Ltd, Espoo, Finland
Elias Hakalehto, Finnoflag Oy, Kuopio and Siilinjärvi, Finland; Department of Agricultural Sciences, University of Helsinki, Helsinki, Finland; University of Eastern Finland, Kuopio, Finland

https://doi.org/10.1515/9783110724967-003

Heat treatment and other traditional preservation techniques have retained their importance among new emerging techniques. Physical methods aim to inhibit, destroy, or remove harmful microbes without additives. Dehydration is an efficient way to preserve food and different techniques are available, including drying and freeze-drying. In some cases, microbial metabolites, for example, fermentation products, including organic acids like lactic acid, can act as preservative agents. Table 2.2 summarizes examples of hurdles to preserve foods.

Table 2.1: Examples of the drivers for mild processing techniques.

Drivers for mild processing techniques include
– Consumers demand for healthier foods that retain their original nutritional properties
– The shift to ready-to-eat and convenience foods that require less further processing by consumers
– Consumer preference for more natural foods that require less processing and fewer chemical preservatives
– Long shelf life; this is more important for retailers but also for consumers and it minimizes food waste; no processing is not an alternative

Table 2.2: Examples of hurdles to preserve foods.

Type of hurdle	Examples
Physical hurdles	Aseptic packaging
	Electromagnetic energy (microwave, radio frequency, pulsed magnetic fields, high electric fields)
	High temperatures (blanching, pasteurization, evaporation, sterilization, extrusion, baking, frying)
	Ionizing radiation: gamma rays, X-rays, electron beam
	Low temperatures (chilling, freezing)
	Modified atmosphere
	Packaging films (incl. active packaging, edible coatings)
	Photodynamic inactivation
	Ultra-high pressures
	Ultrasonication
	Ultraviolet radiation

Table 2.2 (continued)

Type of hurdle	Examples
Physicochemical hurdles	Carbon dioxide, hydrogen peroxide
	Ethanol, lactic acid, organic acids, lactoperoxidase, low pH, low redox potential, low water activity, Maillard reaction products, oxygen, ozone, phenols, phosphates, salts, smoking, sodium nitrite/nitrate, sodium and potassium sulfite, spices and herbs, surface treatment agents, chlorine-based rinsing liquids
Microbially derived hurdles	Antibiotics, bacteriocins
	Bacteriophages
	Competitive microbiota, protective cultures, fermentation

Consumer acceptance is linked to consumer education and preference for less treated foods. Safety of the technologies must be validated through research and cooperation with regulatory agencies.

Combination of nonthermal technologies with other preservation techniques, for example, use of additives, has been examined. The hurdle technique aims to combine several mild preservation techniques and thereby preserve foods (Leistner, 2000; Singh and Shalini, 2016). Understanding of the complex interactions of temperature, water activity, pH and other factors in food matrix or processes is used to design a series of hurdles that can ensure microbiological safety of processed foods.

An integrated concept covering the entire food process is needed for the improvement of food safety. For example, bakery industry has successfully adapted clean room technology in their production facilities during recent years. In Finland, Uotilan Leipomo combined aseptic packaging facilities and controlled process air. This approach significantly improved product shelf life, and the company was able to reduce amounts of preservatives in their products (Uotila, 2006).

2.2 Storage stability

Storage stability and shelf life of foods are influenced by several factors including food properties like level of moisture, water activity, free nutrients, product chemical composition, pH, microbial load, packaging type, and storage conditions (e.g., temperature, relative humidity, and exposure to light; Hammond et al., 2015). Three main food spoilage categories include physical spoilage, chemical spoilage (including degradation, oxidation, other enzymatic activities, and Maillard reaction) and microbiological spoilage (Martin et al., 2021). In the marketing of industrial products, the retention of nutritional values is to be determined also after preparation of food products, which often precedes the entry of the products into the market, for example, the convenience foods (Lee, 2018). Then the results from storage of new materials or intermediate products and their stability are evaluated together with the effects of the final processing steps. The storage stability and usability are determined by experiments specifically designed for different food assortments.

2.3 Traceability and other means to maintain the food quality during the distribution chain

Globalization and the growing export and import of goods across borders have increased the need for improved tracking and safety evaluation, especially in food industry where there are highly variable national regulations. This has increased the demand for enabling technologies that can track and share product and production information across regions and value chain in a cost-effective manner.

From an environmental perspective, food production is resource-intensive and has significant environmental impacts. Hence, if food is lost or wasted, this entails poor use of resources and negative environmental impacts. It has been estimated that growing population and rising income will increase the demand of agricultural products by 35–50% between 2012 and 2050, which puts more pressure on food systems. Three types of environmental footprints of food loss and waste are quantifiable: GHG emissions (carbon footprint), pressures on land (land footprint), and pressures on water (water footprint). Strongest impact on these parameters is obtained by combining different interventions, including food loss and waste reduction (FAO, 2019).

Globally, around 14% of food produced is lost starting from the post-harvest stage, and up the food supply chain toward the client, excluding the retail stage. Currently, accurate estimates of waste by retailers and consumers are being prepared (FAO, 2019). Losses of fruits and vegetables are significantly higher than that of cereals and pulses (Figure 2.1). This is linked to their perishable and fragile nature. The causes at retail level are linked to the limited shelf life of perishable products, to private quality standards of buyers and variability of demand, in particular for fresh produce. Storage conditions, packaging quality and handling practices have a significant impact on the quality, shelf life and acceptability of food products.

Consumer preferences for high-quality and "perfect" products also cause food waste (Møller et al., 2016). The consumption stage is a critical waste point for all types of products. High waste percentages have been reported for highly perishable foods like animal products (14–37%) and fruits and vegetables (9–20%). In industrialized countries, as the majority of the wasted food is lost toward the end of the food supply chain, targeting retail or consumer waste may bring about the largest reduction in food loss and waste as well as the environmental damage it causes.

Figure 2.1: Losses of fruits and vegetables are relatively high in the food chain. In the fruit and vegetable production, also the relatively strict quality requirements often cause increasing amounts of discarded products. Fortunately, there are novel potential technologies to fully exploit such rounds as these desert-grown organic peppers as raw materials in the biorefineries. Photo: Elias Hakalehto.

According to Natural Resources Institute Finland, the annual food loss in Finland is 400 M kg. Reducing food loss and waste is an important target of the sustainable development goals (SDGs), as well as a mean to achieve other SDG targets, related to food security in particular, nutrition and environmental sustainability. SDG Target 12.3 calls for halving per capita global food waste at retail and consumer levels by 2030 and reducing food loss (including post-harvest loss) along production and supply chains. Food loss and waste reduction have potential to improve the environmental sustainability of food systems significantly. FAO (2019) estimated that:

- The global carbon footprint of food loss and wastes, excluding emissions from land use change, is 3.3 gigatons of carbon dioxide equivalent, corresponding to about 7% of total GHG emissions.
- The use of surface and groundwater resources (blue water) attributable to food loss or waste is about 250 km^3, representing around 6% of total water withdrawals.
- Almost 1.4 billion hectares, equal to about 30% of the world's agricultural land, are used to produce food that is later lost or wasted.

In Europe, we are mostly privileged in comparison with many other areas in having both food security and strict food safety requirements. Food supply chain integrity is a complex and multifaceted concept (Davidson et al., 2017). It includes food safety, security, traceability, origin authenticity, quality attribute and product information resulting in final food product integrity. In the EU, food industry (producers, distributors, wholesalers, and retailers) has the responsibility for ensuring that retail foodstuffs are safe for human and animal consumption (EU legislation). In Europe, according to the General Food Law Regulation, traceability is defined as the ability to trace and follow food, feed, and ingredients through all stages of production, processing, and distribution. IFS Food Standard (for auditing quality and food safety of food products, IFS 214) and Global Standard Food Safety (BRC, 2015) are also available for adoption by food industry and to complement the European Union (EU) legislation.

Food safety, diminishing of food waste, and consumers need for reliable information are important through the food ecosystem. Traceability and unbroken logistic chain safeguard high-quality and safe products for the consumers. In case of perishable food, for example, temperature tracing is mandatory. In some cases, carbon dioxide and moisture are also wildly tracked. Biorefineries should have an increasing role in the future food ecosystem (Hakalehto, 2020).

Traceability systems ensure quality of raw materials, allowing certification and accreditation of the products, providing means for quick action, for example, in cases of withdrawals, location of the products, implementing control systems, preventing fraud, and anti-tampering of the goods (Espiñeira and Santaclara, 2016). Traceability and development of new technologies are important, and these issues have been studied, for example, in the EU-funded projects like TagItSmart (Figure 2.3, https://www.tagitsmart.eu/). Also FAO (2017) has created guidance for traceability).

Figure 2.2: Examples of factors affecting on the food quality and process hygiene.

2.4 Proper handling of foodstuffs

In EU, the food hygiene proficiency is regulated by the EU's General Regulation on Food Hygiene (Regulation EC No. 852/2004) and on national level, for example in Finland, by the Food Act 297/2021: https://www.finlex.fi/fi/laki/alkup/2021/20210297.

A hygiene passport system is utilized in Finland for the demonstration of food hygiene proficiency. According to the system, a person has to have Hygiene Passport if they work on food premises or handle unpacked easily perishable foodstuffs. Food hygiene proficiency test includes elements from microbiology, food poisoning, hygienic working methods, personal hygiene, cleaning, in-house control, and legislation. In Finland, the Finnish Food Authority keeps register on authorized hygiene passport examiners. The official regulations regarding food hygiene are exemplified in Table 2.3.

Table 2.3: Examples of relevant principles of the hygiene rules according to the European Union (Regulation EC No. 852/2004: https://eur-lex.europa.eu/legal-content/EN/TXT/?uri=CEL EX:02004R0852-20090420).

Examples of relevant principles of the hygiene rules according to the European Union
– Primary responsibility for food safety on the food business operator
– Food safety ensured throughout the food chain, starting with primary production – from farm to fork
– General implementation of procedures based on the Hazard Analysis and Critical Control Points principles (HACCP)
– Application of basic common hygiene requirements, possibly further specified for certain categories of food
– Registration or approval for certain food establishments
– Development of guidelines to good practice for hygiene or for the application of HACCP principles as a valuable instrument to aid food business operators at all levels of the food chain to comply with the new rules
– Flexibility provided for food produced in remote areas (high mountains, remote islands) and for traditional production and methods

A food business operator must ensure that foodstuffs are safe, and the quality and the composition of the foodstuffs are compliant with the legislations. In addition, the information provided to the customer must be truthful, sufficient, and should not mislead the consumer by any means, for example, appropriate labeling needs to be provided.

Several factors including process air quality, process surface quality, raw material quality, personnel hygiene proficiency, and good manufacturing practices influence on the end-product quality and storage stability (Figure 2.2).

Packaging is part of the food chain integrity, and packaging materials need to be food grade. It needs to comply with chemical safety (incl. no harmful substances),

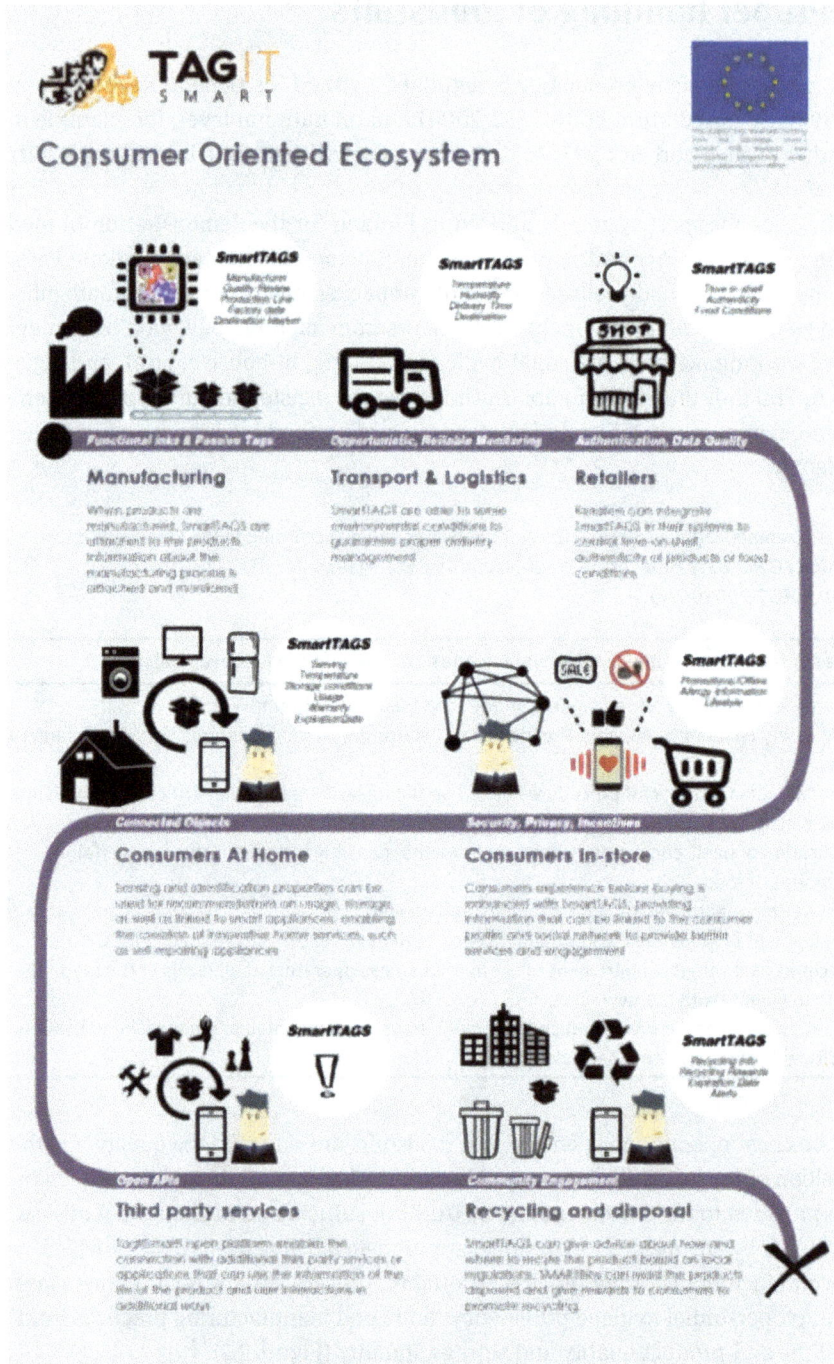

Figure 2.3: An example of utilization of smart tags in the food chain. Example created by the TagItSmart project (https://www.tagitsmart.eu/D1.1.pdf).

microbiological safety (microbiological quality of the end product), and the sensory quality (no off-flavors or volatiles). Proper packaging protects the product and increases shelf life. Sensor equipped, intelligent package systems have been developed for keeping the hygienic qualities of foods (Fuertes et al., 2016). Also, smart trays have been developed for quality follow-up.

2.5 Critical points in maintaining the food quality during pandemics

Based on observations and experiences during recent COVID-19 pandemic, special attention should be paid to the following issues during pandemics:

– As a preparedness for quarantine and limited movement situations, households should have sufficient amount of food in emergency stock. Food with long shelf life is important in this case.
– During quarantine, food delivery services are important. Quality of the distribution chain and hygiene practices are important.
– Global logistics may have delays. Food with long shelf life is important, in this case, for disturbed food services.
– Limitations in restaurant services have emphasized supply and demand of take-away food. Besides challenges in maintaining the food quality during the distribution chain (incl. cold chain), the shortage of packaging material can be a problem.
– Sudden shortage of various items is possible and likely in long term. For instance, lots of reagents and laboratory materials are used for performing COVID-19 analyses, which may have an influence on the availability of materials needed for regular food quality analysis.

2.6 Conclusions

Although the norms and legislation mostly give clear messages about the dealing with various food commodities, it is always possible that changing conditions, variation in food availability, or other factors make it necessary to change the practices either temporarily or for good. For example, some food components can turn out to be unsuitable for human consumption or the food sortiment needs to be processed in an alternative manner. It is important to have up-to-date laws and regulations in the food sector to protect the consumers. In addition to the national guidelines, also such organizations as European Union and FAO set their goals for monitoring food quality. Any major alteration in the public health or epidemics situation should initiate immediate revision of the rules, if necessary. Extensive food research is essential, as well as the international communication and exchange of ideas. Globally numerous conferences or currently webinars, are organized on these topics, starting from the food manufacturing and technologies to the methods to be applied for the understanding of the food chains in the societies. For example, the US Federal Government is subventing annual international food chemical technology conferences (Food Chemical Technology conferences by United Scientific Inc.). Such gatherings are popular meeting points for the food sector professionals and organizations to establish contacts between university researchers, companies, and authorities. The ultimate aim of these joint activities is to help the societies to maintain healthy, safe, sufficient and sustainable food services at all times.

References

BRC. (2015). Global Standard Food Safety. The British Retail Consortium. SAI, Global, Milton Keynes.

Davidson, R., Antunez, W., Madslien, E., Belenguer, J., Gerevini, M., Torroba, T., Prugger, R. (2017). From food defence to food supply chain integrity. British Food Journal, 1: 52–66.

Espiñeira, M. and Santaclara, F.J. (2016). What is food traceability? In: Espiñeira, M., Santaclara, F.J. (eds.). Woodhead Publishing Series in Food Science, Technology and Nutrition, Advances in Food Traceability Techniques and Technologies. Woodhead Publishing: 3–8.

FAO. (2017). Food traceability guidance. http://www.fao.org/3/i7665e/i7665e.pdf.

FAO. (2019). The State of Food and Agriculture 2019. Moving forward on food loss and waste reduction. Rome. Licence: CC BY-NC-SA 3.0 IGO. http://www.fao.org/3/ca6030en/ca6030en.pdf.

Fuertes, G., Soto, I., Carrasco, R., Vargas, R., Sabattin, J., Lagos, C. (2016). Intelligent packaging systems: sensors and nanosensors to monitor food quality and safety. Journal of Sensors, Article ID 4046061. https://doi.org/10.1155/2016/4046061.

Hakalehto, E. (2020). Chicken IgY antibodies provide mucosal barrier against SARS-CoV-2 virus and other pathogens. IMAJ 23 (4): 208–211.

Hammond, S.T., Brown, J.H., Burger, J.R., Flanagan, T.P., Fristoe, T.S., Mercado-Silva, N., Nekola, J.C., Okie, J.G. (2015). Food spoilage, storage, and transport: Implications for a sustainable future. BioScience, 65 (8): 758–768, https://doi.org/10.1093/biosci/biv081.

Lee, K.C.L. (2018). Grocery shopping, food waste, and the retail landscape of cities: The case of Seoul. Journal of Cleaner Production, 172: 325–334.

Leistner, L. (2000). Basic aspects of food preservation by hurdle technology. International Journal of Food Microbiology, 55: 181–186

Martin, N.H., Torres-Frenzel, P., Wiedmann, M. (2021). Invited review: Controlling dairy product spoilage to reduce food loss and waste. Journal of Dairy Science, 104 (2): 1251–1261. doi: 10.3168/jds.2020-19130. Epub 2020 Dec 11. PMID: 33309352.

Møller, H., Hagtvedt, T., Lødrup, N., Andersen, J.K., Lundquist Madsen, P., Werge, M., Kirstine Aare, A., Reinikainen, A., Rosengren, Å., Kjellén, J., Stenmarck, Å, Youhanan, L. (2016). Food waste and date labelling Issues affecting the durability. TemaNord 2016: 523. Nordic Council of Ministers 2016. https://norden.diva-portal.org/smash/get/diva2:950731/FULLTEXT04.pdf

Singh S, Shalini R. (2016). Effect of Hurdle Technology in Food Preservation: A Review. Critical Reviews in Food Science and Nutrition, 56 (4): 641–649. doi: 10.1080/10408398.2012.761594. PMID: 25222150.

Uotila, P. (2006). Cleanroom technology application in bakery. In: Wirtanen, G. and Salo, S. (eds.). 37th R3 - Nordic Contamination Control Symposium. VTT Symposium 240: 129–131. https://www.vttresearch.com/sites/default/files/pdf/symposiums/2006/S240.pdf.

Frank Adusei-Mensah, Elias Hakalehto, Carina Tikkanen-Kaukanen

3 Microbiological and chemical safety of African herbal and natural products

Abstract: Products of natural origin have been used globally for centuries for food and medicine. In recent years, consumption of herbal medicines, functional foods, cosmetics, and other non-therapeutic products of natural base has become popular. Cultivation, manufacturing, transportation, and storage conditions of natural and herbal products expose them to the risk of microbial and chemical contamination. Pesticide, heavy metal, and microbial contamination of herbal medicinal products have been associated with numerous pathogenic conditions in humans with great public health concerns. In order to monitor the standard operating procedures, regular evaluation is the key to ensure patient safety.

3.1 Background

Various products of natural origin play an instrumental role in human existence on the Earth, and their usage dates back to centuries. Herbal products have been used as medicine before the pharmaceutical medicines. Herbal medicine forms an essential part of "traditional medicine" (TM) also recognized as complementary and alternative medicine (WHO, 2007). According to the European Commission, herbal medicinal products (HMPs) are defined as any medicinal product exclusively containing one or more herbal substances as active ingredients, one or more herbal preparations, or a combination of the two (European Commission, 2004). Similarly, the Finnish Medicines Agency (FIMEA) also defines HMPs as medicinal products (including supplements) that contain herbal substances, herbal preparations, or a combination of these as their active agents (FIMEA, 2004). In the last decades, there has been cumulative growth in the usage and popularity of over-the-counter (OTC) plant-based medicinal products, functional foods, and therapeutic products from other natural sources in both developing and developed countries (Adusei-Mensah and Inkum, 2015, Adusei-Mensah et al., 2019a). Demand for organic and natural

Frank Adusei-Mensah, Finnoflag Oy, Kuopio, Finland; Department of Public Health, Institute of Public Health and Clinical Nutrition, University of Eastern Finland, Kuopio, Finland
Elias Hakalehto, Finnoflag Oy, Kuopio, Finland; Department of Agricultural Sciences, University of Helsinki, Finland
Carina Tikkanen-Kaukanen, University of Helsinki, Ruralia Institute and Helsinki Institute of Sustainability Science, Finnish Organic Research Institute, Mikkeli, Finland

https://doi.org/10.1515/9783110724967-004

products in general and plant-based products in particular in recent years for numerous purposes has grown. The increasing resistance to pharmaceutical medicines, lack of definitive and cost-effective therapies for some diseases (Lantto, 2017), adverse health events, and less accessibility and affordability of pharmaceutical medicines in parts of developing and underdeveloped countries have strengthened dependency on natural and plant-based medicines. Furthermore, natural and the mostly organic nature of these products are worth mentioning. Currently, billions of the current global population including close to 1.0 billion Africans (over 70%) use such systems for their primary health-care needs (WHO, 2002). Such surge in acceptance has also come with concerns and fears over the quality, efficacy, and safety of the natural and plant-based products on the market, especially during and beyond the coronavirus pandemic. Natural and plant-based products may be contaminated with microbial contaminants, excessive use or banned pesticides, heavy metals, chemical toxins, and adulteration with pharmaceutical drugs (Adusei-Mensah et al., 2019a, b). Cultivation, harvesting, transporting, manufacturing, and preservation of the natural and/or plant-based products are critical points for microbial and chemical contamination (Kosalec et al., 2009). This assertion and the critical points again highlight the importance of continual research to guide and support the development and distribution of natural and plant-based medicines and practices to provide suitable, safe, and effective products (de Sousa Lima et al., 2020). Implementing and performing the hazard analysis of critical control point (HACCP) on the standard operating procedures leading to the good agricultural practices, the good laboratory practice, the good supply practice, and the good manufacturing practice for safe production of natural and herbal products to ascertain consumer safety is vital (Chan, 2003). The general safety of the products is a necessity for these products to be used as complementary medications or as a part of nutritional substances.

Mineral mining and the increasing use of pesticides for cultivation, post-harvest storage, handling, and production processes exposes some natural and medicinal plant materials and their products to possible heavy metal and pesticide contamination, microbial and microbial-toxin infestation posing health risk. Adverse health risks due to overexposure to chemicals such as minerals and heavy metals including mercury (Hg), lead (Pb), cadmium (Cd), arsenic (As), copper (Cr), and pesticides (Adusei-Mensah et al., 2018; Adusei-Mensah, 2020) have been reported. Again, herbal and natural products may be exposed to microbiological contamination at the field of cultivation, production, and product handling (Figure 3.1). This chapter therefore emphasizes the microbiological and chemical safety of African herbal and natural products and the need for safe practices during and after the coronavirus pandemic.

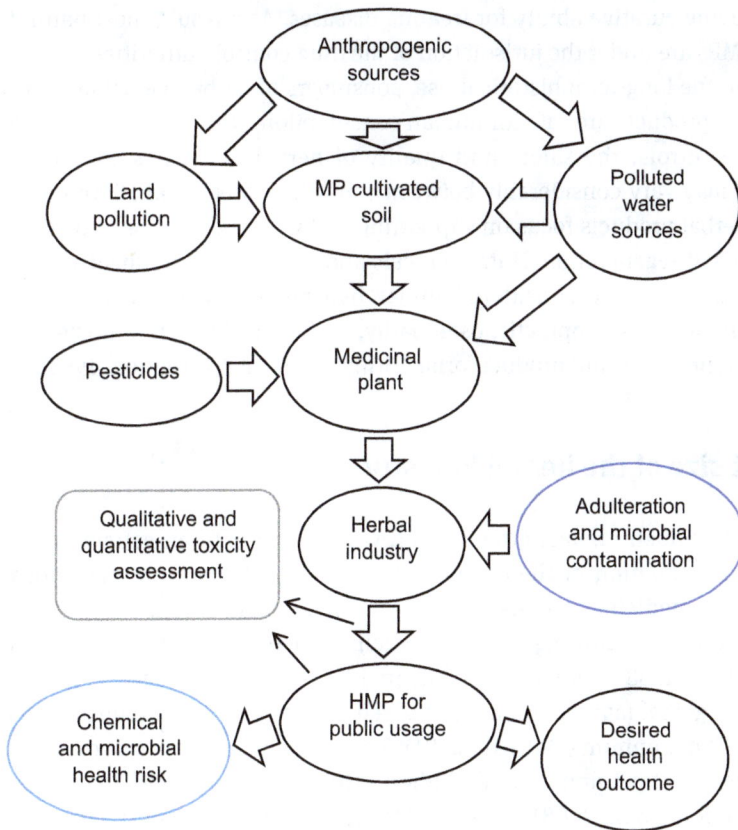

Figure 3.1: Schematic health risk cycle of plant-based medicinal products. MP, medicinal plants; HMP, herbal medicinal product. Adapted from the dissertation of Adusei-Mensah (2020). Risk points are highlighted in blue.

3.2 Herbal and natural products

Amidst the global acceptability of herbal and natural products, their regulatory policies vary among different countries globally. The regulation of herbal medicines in the African region is faced with numerous complexities due to multiple critical points of microbial and chemical contamination, as well as instrumental and resource constrains faced by some member states. Despite the hurdles, however, many governments have taken steps and political will to regulate natural and plant-based products in their region and to protect the health of consumers. As such, steps such as drafting of guidelines for the registration and marketing of HMPs and related substances have been carried out in different countries across the region. In Ghana and in most African countries, HMPs are considered as drugs and they are advertised as having curative ability for treating diseases. As a result, most natural products and HMPs are under the jurisdiction of the drug control authorities.

Grounded on the long-term historical use, consumers of herbal medicines have the belief that the products are safe for human consumption. However, without regular regulatory controls, the safety and quality of herbal medicines cannot be guaranteed and may vary considerably between countries (Street et al., 2008). The proponents of herbal products focus on supporting and working with the body's innate abilities for self-regeneration. Nutraceuticals and plant-based products and alternative medicines may be considered under five broad and distinct clinical disciplines: acupuncture, chiropractic, osteopathy, herbal medicine, and homeopathy with different practices and product forms (British Medical Association, 1993).

3.2.1 Market size of the herbal industry

Herbal medicinal industry is fast growing and widespread with an estimation of over 70% of the population of the developing and about 65% of the developed world using herbal medicines and natural products. Patients' annual expenditure on traditional systems including herbal medicines in the USA increased from 200 million USD in 1988 to 5.1 billion USD in 1997 (Eisenberg et al., 1998). In the year 2001, the global market of the plant-based medicinal product industry was 16–20 billion USD per annum (Mahady, 2001). The global market grew over 300% from 16–20 billion USD per annum to 60 billion USD per annum in less than a decade (Tilburt and Kaptchuk, 2008). In South Africa, the industry is considered as a multi-billion South African ran industry. In Ghana, the industry is worth billions of Ghana cedis and in Nigeria billions of Nigerian naira. The trend of usage may be positively impacted during and after coronavirus pandemic. The global usage calls for global consideration (Komulainen, 2017). Natural products like "inkivääri shotti" (ginger shot, from ginger, lemon, and honey) and honey-based products have gained some popularity in Finland during the corona pandemic period, with similar trend evident globally. There are unlimited evidence of the use of natural and herbal medicines for the prevention and control of coronavirus infection (Huang et al., 2020).

3.2.2 Herbal medicines usage and tradition in Africa

The culture, affordability, easy accessibility, and year-round availability to the rural population makes natural and HMs a reliable source for the treatment of common ailments in Africa (Adusei-Mensah et al., 2020). HMs have deep-rooted tradition in many African countries which dates to centuries. The advancement in technological innovations in the last decade is gradually transforming the herbal industry. Such innovations, the desire for organic and natural products, and the HMs' ability to control a wide spectrum of diseases have had positive impact on the industry resulting in substantial increase in patronage during recent years (Adusei-Mensah and Inkum, 2015). In Ghana, patronage of HMs is believed to be higher among patients with malaria and chronic diseases than among the general population both in Ghana and a similar pattern could be observed in different Afirican countries.

3.3 Plant-based anti-infectives

About 40% of the modern drugs and approximately 75% of drugs for infectious diseases are of natural origin. The number of drug-like molecules present in the vast amount of species (plants, fungi, bacteria, marine invertebrates, and insects) is enormous (Samuelsson and Bohlin, 2009). As a source of novel drugs, plants remain grossly understudied and underused, especially in the developed world (Akerele, 1993; WHO, 2007). A World Health Organization (WHO) study has shown that 80% of the world's population relies solely upon medicinal plants as a source of remedies for the treatment of diseases (Akerele, 1993). Natural products of higher plants may be a source of new antimicrobial agents with possibly novel mechanisms of action (Runyoro et al., 2006; Shahidi Bonjar, 2004; Riihinen et al., 2011).

Many plants contain bioactive compounds like polyphenols, alkaloids, terpenoids, and saponins (Alviano and Alviano, 2009; Mgbeahuruike et al., 2018, 2019a) with several different mechanisms against microbial infections. In addition to anti-adhesive (Toivanen et al., 2011; Huttunen et al., 2011, 2016) or antibiotic properties (Obey et al., 2016; Huttunen et al., 2016), plant material may consist immunomodulatory components that in host alter gut microbiota composition (Matziouridou et al., 2016; Heyman-Lindén et al., 2016).

Because of the complicated nature of COVID-19 virus and its ability for constant and quick modifications, vaccine and therapy development has faced unexpected difficulties. Amidst the available vaccines, there is the need for continual preparedness against current and emerging infectious diseases. A combination therapy targeting different viral factors simultaneously could be a valuable tool against COVID-19 virus. Multiple active ingredients with a holistic effect are the strength of plant-based materials and herbal products. They may possess a valuable solution against viral infections and fight against viral or other microbial pandemics in future.

Honey is an extensively studied natural antimicrobial as shown in numerous scientific reports (Mandal and Mandal, 2011; Combarros-Fuertes et al., 2020; Huttunen et al., 2016; Oinaala et al., 2015; Salonen et al., 2017), including honey from Africa (Nweze et al., 2016; Khan et al., 2014; Mokaya et al., 2020). Recently, a meta-analysis revealed activity of honey against viral respiratory infections (Abuelgasim et al., 2020). Honey has been effectively used for varying medicinal purposes because of its remarkable antimicrobial properties, like in wound healing (Al-Waili et al., 2011). In honeys, several effective antimicrobial components have been described, especially large amounts of methylglyoxal (MGO) present in New Zealand Manuka honey (Kwakman et al., 2011). MGO activity has been described against both bacterial and viral infections (Matsunaga et al., 2014). Against respiratory bacterial pathogens, unique antimicrobial properties of Finnish honeys have been reported (Huttunen et al., 2013; Salonen et al., 2017). Detecting immunostimulatory effects of honeys in gut systems is a novel approach. Immunostimulatory effects of honey have been found from heated honey samples both in vivo and in vitro systems (Ota et al., 2019). Glycoproteins and glycopeptides of Ziziphus honey has been reported to stimulate human neutrophils and murine macrophage cell lines in vitro (Mesaik et al., 2015). Honey saccharides (Ota et al., 2019; Gannabathula et al., 2012), apigenin and kaempferol flavonoids (Majtan et al., 2013), and proteins (Mesaik et al., 2015; Majtan et al., 2006) induce cell-mediated immune responses. Immune modulatory effect of polyphenols has been reported. They regulate immune cell populations, cytokine production, and inactivation of nuclear factor kappa-light-chain-enhancer of activated B-cells (Yahfoufi et al., 2018).

3.4 Medicinal plants against COVID-19 and other epidemics in Ghana

About 80% of the rural population in sub-Saharan Africa depends on traditional herbal remedies for primary health-care and veterinary use (WHO, 2002). In rural parts of China, India, Africa, and Latin America, modern drugs are not available, or, if they are, they often prove to be too expensive, unavailable, or inaccessible. In many developing countries, antibiotic resistance, adverse drug reactions, and the high costs of antimicrobials have made management of infectious diseases ineffective (Kapil 2005; Runyoro et al., 2006). On the other hand, the spread of the virus infections and their bacterial or fungal complications deserve immediate attention. One potential solution could be based on chicken egg yolk antibodies (Hakalehto, 2021). This approach is using the egg yolk of vaccinated hen eggs for prophylaxis by so-called passive immunization. Such procedures with antibodies produced into eggs or colostrum could offer means for preventing disease, together with safe plant production and herbal medications. Neither of the methods is replacing vaccinations as a long-term solution, but provide ways to react locally to the pandemic urgencies.

Domestic chicken forms an important and diverse source of nutrition in rural areas everywhere in Africa. In many cases, the hygienic quality of the chicken feed is compromised (Hakalehto et al., 2014). Chicken and their eggs constitute a basic solution for protein and vitamin acquisition, as well as for many side products (Hakalehto, 2015). Their extensive or industrial use is also bringing about the issues of chicken litter. Poor waste management could worsen the hygienic levels further, unless the manure, parts of the carcasses, and other leftover fractions are properly converted into useful chemicals, energy gases, or other products (Schwede et al., 2017). The dissemination of the viral or other epidemics, in turn, could be prevented by high levels of chemical and microbiological hygiene, and waste recycling. Therefore, the safe usage of various herbal and natural plants or their seeds as chicken feed in the African rural areas should be considered as a part of the holistic picture on the public health.

In Africa, plants are widely used in folkloric medicine and searched as source for novel anti-infectives (Obey et al., 2016, 2018; Mgbeahuruike et al., 2019b). Ethno-pharmacological studies on some indigenous Ghanaian medicinal plants including *Moringa oleifera*, *Paullinia pinnata*, *Sutherlandia frutescens*, *Solanum torvum*, *Alstonia boonei*, *Hibiscus* sp., *Azadirachta indica*, and *Acacia kameruneensis* have shown their efficacy in treating numerous viral infections such as *AIDS*, common cold, pneumonia, and hepatitis (Koffuor et al., 2014; Boadu and Asase, 2017; Firempong, 2020). *Hibiscus* flower is also commonly used in Ghana for a popular local delicacy drink "soobolo." In Kenya, the plant *Croton macrostachyus* has been traditionally used to treat diseases (Wagate et al., 2010). Scientific investigations have shown that ethyl acetate, methanol, and isobutanol stem bark extracts of *C. macrostachyus* are active against several human pathogenic bacteria and have significant antiplasmodial activity against *Plasmodium berghei* causing malaria (Obey et al., 2016, 2018). Mgbeahuruike et al. (2018) have demonstrated that *Piper guineense* from West African TM contains antibacterial alkaloids that could be relevant for the discovery of new natural antibiotics.

The traditional use of herbal plants and products against COVID-19 could be based on their immunomodulating properties, the plants' ability to prevent or terminate viral replication (Firempong, 2020),and their general antiviral properties (Koffuor et al., 2014; Boadu and Asase, 2017). One of the most promising herbal product candidate against COVID-19 infection recommended by Centre for Plant Medicine Research (CPMR) as an immune booster against COVID-19 is the Centre of Awareness Food Supplement (COA FS) (CPMR, 2020). *Hibiscus* sp., *A. indica*, *P. pinnata*,and *S. torvum* are used in the preparation of immune booster in Ghana. COA FS is a multi-herbal commercial product used in Ghana as an immune booster. The aerial parts of *Hibiscus* sp. have been reported to contain anthocyanins, flavonoids, and polyphenols which act as antioxidants, anti-inflammatory agents with cytotoxic activities (Al-Yousef et al., 2020). Its flower infusion is used for treating bronchial catarrh (Siddiqui et al., 2018). *A. indica* leaves contain azadiractin, which has been reported to have impact against the dengue virus (Parida et al., 2002). Azadiractin acts through different modes of mechanisms including inhibiting viral attachment, virucidal, and intracellular inhibition (Xu et al., 2012). Based on molecular docking studies, desacetylgedunin from the seeds of *A. indica* has been reported to be a good drug candidate against COVID-19 (Baildya et al., 2020; Borkotoky and Banerjee, 2020). Andrographolide from *P. pinnata* has been reported to have antiviral activity (Verma, 2020) and quercetin and kaempferol from *M. oleifera* blocks viral replication (Verma, 2020). Torvanol-A and torvanol-H from *S. torvum* have antiviral effect against *Herpes simplex* virus (HSV-I) (Ikeda et al., 2000).

3.5 Safety of natural and herbal medicinal products

The herbal industry has been faced with concerns about pesticides, microbial contamination, and adulteration with pharmaceutical drugs. Those who adulterate do so with the intention of increasing the efficacy and usually without clinical data on compatibility and drug–herb interactions (Adusei-Mensah et al., 2020). Several herbal medicine consumers purchase the herbal medicines OTC without a prescription thereby increasing the patients' health risk. Natural products and herbal medicine consumers could experience adverse health events due to inherent active compounds, microbial contamination, microbial toxins like aflatoxin, pesticide, and heavy metal contamination. Previous research has identified some adverse reactions associated with the use of medicinal herbs. Aristolochic acid from *Aristolochia* species has been reported as being carcinogenic, mutagenic, hepatotoxic, and causes severe toxic effects on kidney (Ekor, 2014). Products containing *Ephedra sinica* have been associated with severe cardiovascular and central nervous system problems (Deng et al., 2013), neurotoxicity, hepatotoxicity, and reversible blindness (Schoepfer et al., 2007). *Piper methysticum* is traditionally used in medications for its anxiolytic activity, however, hepatotoxicity has been ascribed to its use. In addition, these products could pose potential microbial and chemical safety concerns.

3.5.1 Regulation of herbal medicines in Africa

As in many developed countries, most African countries have well established regulatory departments and guidelines for herbal and natural products, though with varying levels of enforcement of these policies. Ghana Food and Drug Authority (GFDA) regulates the production, registration and distribution, and exercise post-registration control of food and drugs in the country. In addition, the GFDA regulates natural products, herbal and alternative medicines in the country. The Nigerian regulatory authority for natural and plant-based products is the National Agency for Food and Drug Administration and Control (NAFDAC). Under the GFDA and NAFDAC, large-scale production, marketing, and post-market evaluation of homeopathic and HMPs require registration and marketing approval. The authorities also regulate post-registration evaluation/monitoring with the intent of protecting the health of the public (Adusei-Mensah, 2020; Osuide, 2002). For safe products and patient safety, the producers and distributors need to take actions to prevent, minimize, control, and report products related health concerns such as microbial and chemical contaminations and adverse health events of natural and herbal medicines. Data on the potential toxicity and safety of many multi-herbal preparations available on the market have not been well established (Adusei-Mensah, 2020; Adusei-Mensah et al., 2019b).

3.6 Microbiological safety of African herbal and natural products

Another major area of health consideration on natural and plant-based products is the microbiological and chemical safety. A plethora of ecological, agronomic, harvesting, manufacturing, processing and storage factors are worth considering for microbial and chemical safety (Street et al., 2008). Risk assessment of the microbial load of natural and plant-based medicines has therefore become an important subject in current HACCP schemes. In this area, considerable amount of previous research on natural and herbal product contamination by microbial organisms, fungi, and microbial toxins have been done. The environments in which the products are produced, transported, and stored make them predisposed to bacterial and fungal contamination. Such contaminations could pose potential health hazards to consumers and economic losses to stakeholders (Figure 3.2). Microbial safety is of critical consideration during and beyond the coronavirus pandemic where the production and transport chain have been hugely impacted. Of major concern is the conceivable presence of pathogenic bacterial, fungi and microbial toxins (Katerere et al., 2008). In a South African study, 16 HMP samples were examined for microbial contamination (Katerere et al., 2008). The microbial isolates were found in 15 of the analyzed samples containing *Aspergillus*, *Fusarium*, or *Penicillium* classes. The mycotoxin Fumonisin B_1 producing plant pathogen was present in 13 of the 16 samples in quantities ranging from 14 to 139 µg/kg (Katerere et al., 2008). This observation suggests the broken microbial quality control between cultivation and the final natural product for patient administration. Microbial cycle from the cultivation field to the final consumer of the natural product has been identified, (Figure 3.2), (Hakalehto et al., 2014).

Microbial coliform forming units (CFU) were observed in 15 traditional medicinal plant samples sold at the Faraday Muthi market, Johannesburg, South Africa (Van Vuuren et al., 2014). Of the observed microbial types, 13% were predominantly opportunistic pathogens and of great health concern. The other detected contaminating microbes were generally non-pathogenic ones with *Pantoea* sp. and *Bacillus* spp. forming a major portion of the isolates. The microbes were mostly within the maximum acceptable contamination limits set by the WHO (i.e., ≤10^5 cfu/g) (Van Vuuren et al., 2014). Infective bacteria such as *Salmonella* spp., *Staphylococcus aureus*, and *Shigella* spp. pose serious health hazards to consumers. Pathogenic microbes and microbial toxins could cause serious health implications. Mycotoxin, for example, aflatoxin, has carcinogenic, mutagenic, teratogenic, neurotoxic, nephrotoxic, and immunosuppressive effect (Turkson et al., 2020).

A study on commercial herbal products from the Nelson Mandela Metropole (Port Elizabeth), South Africa showed significant bacteria and fungi contamination. Different indicator microbes were identified that typify different stages of product contamination. *Bacillus* spp., *Enterobacteriaceae*, *Penicillium*, *Mucor*, and *Aspergillus* were common isolates in the investigated products (Katerere et al., 2008). Cultivation, transportation, and/or harvesting related contamination including pathogenic and non-pathogenic species such as *Acinetobacter baumannii*, *Bordetella* spp., *Chryseomonas* spp., *Flavimonas* spp., *Rahnella aquatilis*, *Pseudomonas* spp., *Stenotrophomonas maltophilia*, *Salmonella* spp., and *Klebsiella pneumoniae* were isolated (Katerere et al., 2008; Govender et al., 2006). The isolation of *Bacillus diarrhea* enterotoxin and multidrug methicillin and vancomycin-resistant *S. aureus* strains in some of the commercial HMPs is of great safety concern. There is, therefore, a call for strict regulatory guidelines for the microbiological quality control of natural and herbal medicines for public safety (Govender et al., 2006). The identified human pathogenic and opportunistic pathogens including gram-negative *Klebsiella ornithinolytica*, *Salmonella*, and *K. pneumoniae* could be a serious public health hazard. Some of these species have been linked to enteric fever, diarrhea, urinary tract infection (UTI), and renal cysts in humans (Famewo et al., 2016).

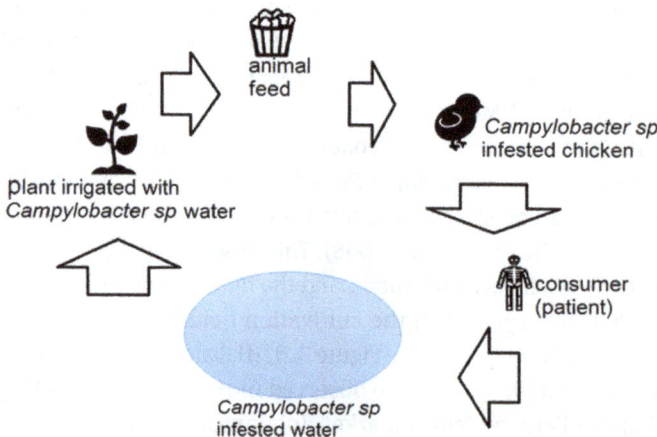

Figure 3.2: Hypothetical route for contamination of *Campylobacter* sp. during the outbreak in Burkina Faso. The route of contamination shows the microbial cycle from the cultivation field to the final consumer of the natural product (modified from Hakalehto et al., 2014).

Bacterial contamination in Malawian commercial herbal medicines from Limbe, Mibawa, and Lunzu markets of Blanyre City was observed by Kalumbi et al., 2020. A total of 68.9% ($n = 29$) of the herbal products showed *Bacillus, Staphylococcus, Klebsiella, Enterobacter*, and *Citrobacter* species and other coliforms, with predominant coliform contamination of 75% of the microbial isolates (Kalumbi et al., 2020). The mean units (CFU) were higher in liquid herbal medicines ($1.4–1.793$) $\times 10^3$ cfu/mL compared to powder ($1.375–1.4$) $\times 10^3$ cfu/mL and tablets ($0.00–0.75$) $\times 10^3$ cfu/mL (Kalumbi et al., 2020). Coliforms are hygienic indicators and the high coliform count in the herbal samples indicates unhygienic practices related to the handling of these products. In herbal products from Ghana, both pathogenic and non-pathogenic microbes were detected. Majority of the studied herbal preparations showed the presence of pathogenic bacteria (Ampofo et al., 2012); however, most of the values were within the acceptable reference limits.

Major pathogenic multidrug resistant (MDR) isolates from HMPs sold in the Gondar Town, Northwest Ethiopia have been reported (Yesuf et al., 2016). A total of 55 HMPs applied locally, orally, or intranasally were analyzed. Fecal indicator organisms (pathogenic and non-pathogenic) including *Escherichia coli, Salmonella, Shigella, S. aureus*, or *Pseudomonas aeruginosa* in 23/55 (41.8%) of the herbal products were detected. The average of total aerobic bacterial count was 1.99×10^8 cfu/ mL, and the average of 1.05×10^8 cfu/mL was reported for the total coliform count. All the isolated aerobic microbes had levels above the WHO tolerable limit and of safety concerns. Contamination in liquid products was also high and follows the trend of other studies (Van Vuuren et al., 2014; de Sousa Lima et al., 2020). *Bacillus* spp., *Enterobacter* spp., *Shigella dysenteriae*, and *Salmonella* spp. were among the most common bacteria isolates reported. These isolates were similar to those found in the studies from South Africa, Ghana, and Brazil (Ampofo et al., 2012; Van Vuuren et al., 2014; de Sousa Lima et al., 2020). Microbes resistant to multiple antibiotic drugs were very common. Most microbial isolates ($n = 150$) were resistant, 131 (87.3%) were resistant to ampicillin, 92 (61.3%) to amoxicillin, 87 (63.3%) to amoxicillin clavulanic acid, and 78 (48.7%) to nitrofurantoin. Pathogenic gram-negative- and gram-positive MDR isolates were resistant to four or more antibiotics (Table 3.1). MDR *Staphylococcus epidermidis* and highly virulent pathogen *Salmonella* spp. with global pathogenic importance were isolated. MDR *Enterobacter cloacae* (9/18), *Providencia stuartii* (4/12), *Bacillus* spp. (8/31), and *Shigella dysenteriae* (3/13) were resistant to four or more antibiotic drugs (Yesuf et al., 2016). Multidrug methicillin and vancomycin-resistant *S. aureus* strains have also been discovered in South Africa (Govender et al., 2006). In Nigeria, microbial isolates resistant to ampicillin, penicillin, co-trimoxazole, and gentamicin antibiotics have been found from herbal products (Adeleye et al., 2005). These findings indicate the growing trend of multiple drug resistance microbes and call the need for policies to regulate non-prescribed antibiotic use in the region.

Table 3.1: Multidrug resistant microbial isolates from natural and herbal medicinal products (according to Yesuf et al., 2016 and Onyemelukwe et al., 2019).

Microbial isolates	Resistant microbial isolates	Antibiotic resistance of the microbial isolates (%)			
	–	MDR2	MDR3	MDR4	≥MDR5
Gram positives (N = 150)	34.7%	7	23.1	17.3	34.6
Gram negatives (N = 150)	65.3%	19.4	22.5	19.4	22.4
Bacillus sp.(N = 40)	6 (15%)	Resistant	–	–	–
E. coli (N = 40)	13 (32.5%)	–	–	–	Resistant
E. asburiae (N = 40)	15 (37.5%)	–	Resistant	–	–
C. krusei (N = 40)	8 (20%)	Resistant	–	–	–

E. asburieae, Enterobacter asburiae; C. krusei, Candida krusei; MDR2, resistance to two antibiotics; MDR3, resistance to three antibiotics; MDR4, resistance to four antibiotics; and ≥MDR5, resistance to five and more antibiotics. Tested antibiotics: co-trimoxazole, gentamycin, tetracycline, nitrofurantoin, amoxicillin clavulanic acid, ceftriaxone, ampicillin, amoxicillin, chloramphenicol, norfloxacin, ciprofloxacin, cloxacillin, penicillin, erythromycin and vancomycin, streptomycin, sparfloxacin, perfloxacin, ofloxacin, cefuroxime, combination of ampicillin and cloxacillin, and rifampicin. N, total number of isolates

In several analyses of the HMPs, most of the microbial isolates detected have been human pathogens (Katerere et al., 2008; Govender et al., 2006; Ampofo et al., 2012; Kalumbi et al., 2020; Famewo et al., 2016) (Table 3.2). *A. baumannii* is an opportunistic nosocomial human pathogen, and is one of the six most important MDR strains globally. *A. baumannii* causes ventilator-associated pneumonia and bloodstream infections with 35% mortality rates in humans. *S. aureus* is a virulent pathogen and is responsible for most hospitalized patients' infections. *Staphylococcus saprophyticus* is a causative agent of uncomplicated lower UTI in humans. Higher contamination of the homemade herbal medicinal plant materials compared to the commercial final products has been reported (Van Vuuren et al., 2014).

Table 3.2: Microbial isolates from herbal and other natural products (according to Katerere et al., 2008; Govender et al., 2006), Ampofo et al., 2012; Kalumbi et al., 2020; Famewo et al., 2016; Onyemelukwe et al., 2019).

Microbe	Contamination level (cfu/mL), or frequency (%)	Nature of microbes	Origin
Acinetobacter baumannii	4 (2.7 %)	Opportunistic MDR human pathogen	SA, Et
Ascaris lumbricoides	22 (53.7%)	Human pathogen	Ni

Table 3.2 (continued)

Microbe	Contamination level (cfu/mL), or frequency (%)	Nature of microbes	Origin
Bacillus spp.	21 (32%)	Opportunistic pathogens in the hospitalized patients	Gh, Ma, Et, Ni
Bordetella spp.	NA	Strictly human pathogen	SA
Chryseomonas spp.	NA	Rare human pathogen	SA
Clostridium spp.	7–596	Human pathogen	Gh
Citrobacter spp.	12 (8.0%)	Emerging urinary pathogen	Et, Ma
Enterobacter cloacae	18 (12%)	Nosocomial pathogenic infections	Et
Enterobacter aerogenes	4 (2.7%)	Opportunistic human pathogen	Et
Enterobacter spp.	5 (9.1%)	Opportunistic human pathogen	Ma, Ni
Escherichia coli	6.7 (10%)	Human and animal pathogen	Et
Ewingella americana	NA	Rare human pathogens	SA
Flavimonas spp.	NA	Emerging human pathogen	SA
Fumonisin B1	14–139 µg/mL	Plant pathogen	SA
Giardia infestinalis	2 (4.9%)	Human pathogen	Ni
Pantoea spp.	NA	Opportunistic human pathogens	SA
Streptococcus pyogenes	4 (2.7%)	Human pathogen	Et
Staphylococcus aureus	4 (7.3%)–10 (6.7%)	Virulent pathogen	Et
Staphylococcus epidermidis	5 (3.3%)	Nosocomial pathogen	Et
Staphylococcus saprophyticus	2 (1.3%)	Human pathogen	Et
Klebsiella spp.	13 (8.7%)	Normal flora and opportunistic pathogens	SA, Ma, Et
Klebsiella pneumoniae	11 (7.3%)	Opportunistic pathogen	ET, SA
Klebsiella ozaenae	4 (1.7%)	Invasive pathogen	ET
Pasteurella pneumolytica	NA	Opportunistic pathogen	SA

Table 3.2 (continued)

Microbe	Contamination level (cfu/mL), or frequency (%)	Nature of microbes	Origin
Providencia spp.	12 (8.0%)	Opportunistic pathogen	Et
Pseudomonas aeruginosa	8–23	Opportunistic pathogen	Gh, Et
Rahnella aquatilis	NA	Pathogenic	SA
Salmonella spp.	6–18	Pathogenic	SA, Gh, Et
Serratia spp.		Opportunistic pathogens	SA, Et
Shigella dysenteriae	13 (8.7%)	Extremely pathogenic	Et
Stenotrophomonas maltophilia	NA	Extremely pathogenic	SA, Et
Staphylococcus spp.	$0.0–4.4 \times 10^5$	Extremely versatile pathogen	Ma, Ni
Total heterotrophic bacteria	$1–2.32 \times 10^3$ cfu/mL	All are not human pathogens	Gh
Total heterotrophic	118–1648 cfu/mL	Human pathogen	Gh
Toxocara canis	5 (12.2%)	Human pathogen	Ni
TC cfu/100 mL	$47–2.1 \times 10^9$ cfu/g	Purity indicator	Gh, Eth
Total aerobic count	$1.27 \times 10^3–1.19 \times 10^9$	Non-human pathogen	Gh, SA
Mold	$1–1.30 \times 10^3$	Non-human pathogen	Gh
Yeasts and molds	1.09×10^3	Pathogen and non-pathogenic spp.	Gh

MDR, multiple drug resistance; TC, total coliform bacteria count; Et, Ethiopia; Gh, Ghana; Ma, Malawi; Ni, Nigeria; SA, South Africa. NA, not available, WHO safe limits: *E. coli* (10^3), *S. aureus* (10^3), *Enterobacter* (10^3), total aerobic mesophilic count (10^3), yeast and molds (10^3), total aerobic count (10^7 cfu/mL), *Salmonella* spp. (10^2 cfu/mL).

3.7 Chemical safety of African herbal and natural products

3.7.1 Heavy metals and health risk

The presence of contaminants, adulterants, and inherent toxic compounds in herbal medicines has been associated with adverse events of herbal medicines administration (Mosihuzzaman and Choudhary, 2008). Heavy metals are defined as naturally occurring metallic elements which have relatively high atomic weights and a density of at least five times greater than that of water (≥ 5 g/cm^3) (Tchounwou et al., 2012). Essential trace metals form key components of several key enzymes that play important roles in various vital physiological processes (Tchounwou et al., 2012). Trace metals and microminerals including Cobalt (Co), copper (Cu), chromium (Cr), iron (Fe), magnesium (Mg), manganese (Mn), nickel (Ni), and zinc (Zn) are vital minerals required in small quantities for various biochemical and physiological functions (WHO, 2004). Excessive amount of the trace metals or their lack in the human system have been linked to various adverse health effects in human with various pathogenic conditions. Chromium (Cr(IV)) and copper have narrow concentration range for beneficial effect, and outside this narrow range leads to toxic adverse health effects (Achadu et al., 2016).

Vanadium (V) and manganese are essential for proper enzyme functioning, but oxidized vanadium such as vanadium pentoxide (V_2O_5) is carcinogenic and, when inhaled, causes mutagenic damage in the DNA (Tchounwou et al., 2012). Manganese, iron, and nickel have known importance in the body. However, permanganate ion (MnO_4^-) is a known liver and kidney poison and nickel carbonyl ($Ni_2(CO)_4$) is known to cause extreme toxicity in human causing respiratory failure, brain damage, and death (Budavari et al., 1996). Ingesting more than 0.5 g of iron a day can induce cardiac collapse and death (Tchounwou et al., 2012). A high dose of copper sulfate ($Cu(SO_4)_2$) has been reported to cause major organ damage and death.

Table 3.3: Levels of heavy metal content (mg/kg) in selected herbal and natural medicinal products (according to Adusei-Mensah et al., 2019a; Adusei-Mensah, 2020; Kalumbi et al., 2020; Turkson et al., 2020).

Heavy metal	Concentration Range (mg/kg)	MRL	Percentage of samples exceeding acceptable limit	Origin
As	0.2599–1.2847	0.02	30% ($N = 26$)	Ghana, Malawi
Cd	0.0018–2.6000	0.06	8.33% ($N = 26$)	Ghana, Malawi
Cu	1.7549–8.7860	0.1	87.5% ($N = 8$)	Ghana
Cr	0.2919–1.3567	0.05	100% ($N = 6$)	Ghana
Fe	0.0667–26.7210	0.3	50% ($N = 2$)	Ghana
Hg	ND–0.0027	0.01	3.85% ($N = 26$)	Ghana, Malawi
K	3.6900–6.3222	NA	NA ($N = 2$)	Ghana
Mn	0.8438–3.9204	0.26	100% ($N = 8$)	Ghana
Na	0.3700–1.2150	NA	NA ($N = 2$)	Ghana
Ni	0.1729–1.2188	0.6	75% ($N = 8$)	Ghana

Table 3.3 (continued)

Heavy metal	Concentration Range (mg/kg)	MRL	Percentage of samples exceeding acceptable limit	Origin
Pb	0.0234–1.0000	0.1	57.92% ($N = 26$)	Ghana, Malawi
Zn	0.0760–0.4380	27.4	0% ($N = 2$)	Ghana

MRL, maximum residual limit; ND, non-detected; NA, not available; N, total number of samples investigated.

Heavy metal levels above the established maximum residual limits (MRLs) pose a health risk to consumers. In our recent study (Adusei-Mensah et al., 2019a), heavy metal contamination was detected in most of the studied HMPs (HMPA-F) (Table 3.3) at levels above their MRL. Heavy metal introduction may occur during cultivation and or manufacturing. The maximum residual content of Cr, Mn, Ni, Cu, and As were above the MRL in all the six studied herbal preparations (Adusei-Mensah et al., 2019a; Table 3.3). Pb contents were mostly below the MRL. MRLs are important for legal purposes but not decisive for health risk estimation due to differing consumption frequencies, dosage variations, and body mass index variations.

Kalumbi et al. (2020) observed high lead and cadmium contamination of 67% ($n = 18$) and 11% ($n = 18$), respectively, in the HMPs collected from Blantyre, Malawi. The detected levels exceeded the acceptable reference limits. They discovered that liquid herbal products contained the highest concentrations of heavy metals (Kalumbi et al., 2020). The presence of high levels of heavy metals including arsenic is associated with cancer risk. High levels of heavy metals in herbal products could be due to cultivation conditions of the medicinal plants or unhygienic processing or storage conditions.

3.7.2 Pesticides and health risk

According to the Food and Agriculture Organization, pesticides are defined as any substance or mixture of substances intended for preventing, destroying, or controlling any pest. Pesticides have been useful for decades in both the agricultural and health industries. Pesticides provide an efficient, economic, labor, and lifesaving means for pest control in both agriculture and public health sectors (Cooper and Dobson, 2007; Damalas and Eleftherohorinos, 2011). Despite the enormous benefits of pesticides, their extensive use presents health risk to humans and they are of great concerns (Skovgaard et al., 2017). Data on global health impacts of pesticides are limited, but a lot can be learnt from the few reported data. For instance, in 2002 the impact of suicidal pesticide ingestion has been estimated to have caused 186,000 deaths and 4,420,000 disability adjusted life globally (Gyenwali et al.,

2017). It has been reported that over 98% of sprayed insecticides and 95% of herbicides reach non-targeted destinations including other ecologically important species, air, water, and soil, thereby polluting the environment and causing health risk (Tsimbiri et al., 2015). Excess and run-off pesticides seep into the soil and are washed into water bodies and contaminate them. Pesticides with long half-life stay in the environment and end up in the tissues of plants including herbal medicinal plants and into humans via the food chain.

Intake of certain pesticides in amounts above their safe levels can lead to acute or chronic poisoning, coma, and death (Skovgaard et al., 2017). Reduced intelligence quotient power cancer and neurological problems are some of the health problems associated with chronic exposure to some pesticides (Skovgaard et al., 2017). The use of aldrin, camphechlor, chlordecone, dieldrin, endrin, dichlorodiphenyltrichloroethane, heptachlor, and mirex is prohibited or severely restricted by the Stockholm convention (United Nations, 2001). But according to recent reports, some of these banned hazardous pesticides are still being used in many countries including Ghana (Ghana Environmental Protection Agency, 2008).

Table 3.4: Pesticide residues detected and quantified from the herbal products (according to Adusei Mensah et al., 2018).

Sample	Pesticide class	Pesticide	Mean pesticide level (mg/kg)	MRLs (mg/kg)
HMPA	Organophosphorus	Chlorpyrifos	0.022	0.05
HMPB	Organophosphorus	Chlorpyrifos	0.01	0.05
	Organophosphorus	Fenitrothion	0.05	0.1
HMPC	Pirimiphos-M	Pirimiphos-m	0.008	0.05
	Organophosphorus	Chlorpyrifos	0.042	0.05
HMPD	Organophosphorus	Chlorpyrifos	0.037	0.05
HMPE	Organophosphorus	Pirimiphos-m	0.082	0.05
	Organophosphorus	Chlorpyrifos	0.005	0.05
HMPF	Organophosphorus	Chlorpyrifos	0.0175	0.05

HMPA, herbal product A; HMPB, herbal product B; HMPC, herbal product C; HMPD, herbal product D; HMPE, herbal product E; HMPF, herbal product F; MRL, maximum residual limit.

Chlorpyrifos is a commonly used pesticide in Ghana (Adusei-Mensah, 2020). Though its use is legal, care must be exercised during usage to reduce human exposure and health risk. Some legally banned pesticides revised to (G-EPA 2008) including aldrin, dieldrin, and chlordane, were also identified in 50% of the herbal products sampled in Ghana (Adusei Mensah et al., 2018). This may indicate their continues use in the country despite the ban of their usage in the country. Authorities are therefore called on to regulate their importation and usage in the country. In our previous study, we found that the total pesticide content of the studied six (HMPA-F) herbal preparations were mostly (83%, $n = 6$) within internationally acceptable safety limits (Adusei Mensah et al., 2018; Table 3.4). The concentration of pirimiphos-methyl in HMPE was higher than the MRL (Table 3.4) and may predispose to pirimiphos-methyl exposure and possible health hazard following chronic use of this herbal preparation.

3.8 Fraudulent practices and adulteration

Adulteration of food and medicines is becoming a global challenge of the twenty-first century where chemicals like melamine are added to food and dairies to false increase its protein estimation value. Adulteration in the natural and herbal product industry cannot be ruled out. For instance, honey adulteration is becoming a common issue. Honey is usually adulterated with low-cost sugars (Fakhlaei et al., 2020). Honey adulterations include direct, indirect, and blended methods. The direct adulteration is a post-production procedure in which certain foreign sugar syrups and chemical sweeteners are added directly into the honey to increase the quantity and sweetness of the honey (Zábrodská and Vorlová, 2015). During the indirect adulteration process, the honeybees are overfed with industrial sugars to recover more honey from hives (Fakhlaei et al., 2020). In the blending adulteration procedure, pure and high-quality honey is mixed with cheap and low-quality honey. These adulteration procedures increase the possibility of the contamination.

Various analytical methods are applicable for the detection of adulteration in honeys. High-performance liquid chromatography, gas chromatography coupled mass spectrometry, high-performance anion-exchange chromatography coupled pulsed amperometric detection, matrix-assisted laser desorption/ionization mass spectrometry, ultrahigh-performance liquid chromatography coupled quadrupole time-of-flight mass spectrometry, liquid chromatography-electrochemical detection, and other advanced techniques including infrared spectroscopy, nuclear magnetic resonance spectroscopy, and Raman spectroscopy are currently used for honey adulteration analysis (Fakhlaei et al., 2020; Wu et al., 2017).

3.9 Conclusion

The cultivation, harvesting of plant-based or other natural materials, transporting, manufacturing, and preservation of them are critical points for microbial and chemical contamination. As a result, the implementation of high standards and practices will influence the microbial and chemical quality and safety of natural and medicinal products during and beyond the coronavirus pandemic. Regulatory programs for the production, distribution, and storage should be implemented to reduce microbial and chemical contamination and to improve quality. In addition to compliance with safe production and transportation practices, observing current national and international legislations is vital. Among product categories, liquid preparations were most frequently contaminated by microbes. The wet nature of liquid herbal medicinal preparations makes them more vulnerable to microbial contaminations compared to the other product forms. Home-prepared plant-based medicine and natural products have higher risk of contamination by coliforms and pathogenic bacteria than the commercial HMPs. Another observation of great health concern is the increasing trend of MDR microbes. Such findings infer the level of OTC use of antibiotics in the region and a call for concerted effort to control it. For heavy metals, most of the reported metals including arsenic, chromium, copper, lead, and cadmium had levels exceeding regulatory limits. In conclusion, pesticides, heavy metals, and microbial contamination pose a threat to the quality and safety of the plant-based and other natural products and require a concerted effort to ensure their microbial and chemical safety during and beyond coronavirus or other possible pandemics.

References

Abuelgasim, H., Albury, C., Lee, J. (2020). Effectiveness of honey for symptomatic relief in upper respiratory tract infections: A systematic review and meta-analysis. BMJ Evidence-Based Medicine. doi: 10.1136/bmjebm-2020-111336.

Achadu, O., Ochimana, O., Ochefu, A., Njoku, U. P. (2016). Assessment of heavy metals levels and leaching potentials in dumpsites soils in Wukari, North-Eastern Nigeria. International Journal of Modern Analytical and Separation Science, 5: 20–31.

Adeleye, I. A., Okogi, G., Ojo, E. O. (2005). Microbial contamination of herbal preparations in Lagos, Nigeria. Journal of Health, Population and Nutrition, 23: 296–297.

Adusei-Mensah, F. (2020). Toxicological surveillance and safety profile of commonly used herbal medicinal products in Kumasi metropolis of Ghana. PhD Dissertation, University of Eastern Finland.

Adusei-Mensah, F., Inkum, E. I. (2015). Has the recent upsurge in traditional herbal medicine in Ghanaian market been translated into the health of the Ghanaian Public: A retrospective cohort study? IJNRHN, 2: 98–106.

Adusei-Mensah, F., Henneh, T. I., Ekor, M. (2018). Pesticide residue and health risk analysis of six commonly used herbal medicinal products in Kumasi. Texila IJPH, 6: 3.

Adusei-Mensah, F., Essumang, D. K., Agjei, R. O., Kauhanen, J., Tikkanen-Kaukanen, C., Ekor, M. (2019a). Heavy metal content and health risk assessment of commonly patronized herbal medicinal preparations from the Kumasi metropolis of Ghana. Journal of Environmental Health Science and Engineering, 17: 609–661. doi: 10.1007/s40201-019-00373-y.

Adusei-Mensah, F., Haaranen, A., Kauhanen, J., Tikkanen-Kaukanen, C., Henneh, T. I., Ganu, D., Edem Ametsetor, E. S., Olaleye, S. A., Ekor, M. (2019b). Post-market safety and efficacy surveillance of herbal medicinal products from users' perspective: A qualitative semi-structured interview study in Kumasi, Ghana. International Journal of Pharmacy and Pharmaceutical Sciences, 3: 136.

Adusei-Mensah, F., Tikkanen-Kaukanen, C., Kauhanen, J., Henneh, I. T., Owusu Agyei, P. E., Akakpo, P. K., Ekor, M. (2020). Sub-chronic toxicity evaluation of top three commercial herbal antimalarial preparations in the Kumasi metropolis, Ghana. Bioscience Reports, 40: 6.

Akerele, O. (1993). Nature's medicinal bounty: Don't throw it away. World Health Forum. Journal of Pathogens, 14 (4): 390–395. 6.

Alviano, D.S., Alviano, C.S. (2009). Plant extracts: Search for new alternatives to treat microbial diseases. Current Pharmaceutical Biotechnology, 10: 106–121.

Al-Waili, N., Salom, K., Al-Ghamdi, A. A. (2011). Honey for wound healing, ulcers, and burns; data supporting its use in clinical practice. The Scientific World Journal, 11: 766–787. doi: 10.1100/tsw.2011.78.

Al-Yousef, H. M., Hassan, W. H. B., Abdelaziz, S., Amina, M., Adel, R., El-Sayed, M. A. (2020). UPLC-ESI-MS/MS profile and antioxidant, cytotoxic, antidiabetic, and antiobesity activities of the aqueous extracts of three different Hibiscus species. Journal of Chemistry. doi: 10.1155/2020/6749176.

Ampofo, J. A., Andoh, A., Tetteh, W., Bello, M. (2012). Microbiological profile of some Ghanaian herbal preparations – safety issues and implications for the health professions. Open Journal of Medical Microbiology, 2: 720–726.

Baildya, N., Khan, A. A., Ghosh, N. N., Dutta, T., Chattopadhyay, A. P. (2020). Screening of potential drug from Azadirachta Indica (Neem) extracts for SARS-CoV-2: An insight from molecular docking and MD-simulation studies. Journal of Molecular Structure, 129390.

Boadu, A. A., Asase, A. (2017). Documentation of herbal medicines used for the treatment and management of human diseases by some communities in Southern Ghana. Evidence-based Complementary and Alternative Medicine. doi: 10.1155/2017/3043061.

Borkotoky, S., Banerjee, M. (2020). A computational prediction of SARS-CoV-2 structural protein inhibitors from Azadirachta indica (Neem). Journal of Biomolecular Structure and Dynamics, 1–11. doi: 10.1080/07391102.2020.1774419.

British Medical Association. (1993). Complementary Medicine: New Approaches to Good Practice. Oxford: Oxford University Press.

Budavari, S., O'Neil, M. J., Smith, A., Heckelman, P. E., Kennedy, J. F. (1996). Nickel Carbonyl. The Merck Index, 12th. Whitehouse, NJ: Merck & Co, 1028.

Chan, K. (2003). Some aspects of toxic contaminants in herbal medicines. Chemosphere, 52: 1361–1371.

Combarros-Fuertes, P., Fresno, J. M., Estevinho, M. M., Sousa-Pimenta, M., Tornadijo, M. E., Estevinho, L. M. (2020). Honey: Another alternative in the fight against antibiotic-resistant bacteria? Antibiotics (Basel). doi: 10.3390/antibiotics9110774.

Cooper, J., Dobson, H. (2007). The benefits of pesticides to mankind and the environment. Crop Protection (Guildford, Surrey), 26: 1337–1348. doi: 10.1016/j.cropro.2007.03.022.

CPMR. (2020). 9 Herbal Product as immune boosters – CPMR. https://www.cpmr.org.gh/2020/05/16/centre-for-plant-medicine-recommends-9-herbal-product-as-immune-boosters/

Damalas, C. A., Eleftherohorinos, I. G. (2011). Pesticide exposure, safety issues, and risk assessment indicators. International Journal of Environmental Research and Public Health, 8: 1402–1419. doi: 10.3390/ijerph8051402.

de Sousa Lima, C. M., Fujishima, M. A. T., de Paula Lima, B., Mastroianni, P. C., de Sousa, F. F. O., da Silva, J. O. (2020). Microbial contamination in herbal medicines: A serious health hazard to elderly consumers. BMC Complem Altern M, 20 (1): 17.

Deng, Y., Cao, M., Shi, D., Yin, Z., Jia, R., Xu, J., Wang, C., Lv, C., Liang, X., He, C., Yang, Z., Zhao, J. (2013). Toxicological evaluation of neem (Azadirachta indica) oil: Acute and subacute toxicity. Environmental Toxicol Pharmacol, 35: 240–246.

Eisenberg, D. M., Davis, R. B., Ettner, S. L., Appel, S., Wilkey, S., Rompay, M. V., Kessler, R. C. (1998). Trends in alternative medicine use in the United States, 1990–1997: Results of a follow-up national survey. JAMA, 280: 1569–1575.

Ekor, M. (2014). The growing use of herbal medicines: Issues relating to adverse reactions and challenges in monitoring safety. Frontiers Pharmacol, 4. doi: 10.3389/fphar.2013.00177.

European Commission (2004). Herbal medicinal products. available at https://ec.europa.eu/health/human-use/herbal-medicines_en. Visited 28.01.2020.

Fakhlaei, R., Selamat, J., Khatib, A., Razis, A. F. A., Sukor, R., Ahmad, S., Babadi, A. A. (2020). The toxic impact of honey adulteration: A review. Foods (Basel, Switzerland), 9: 11.

Famewo, E. B., Clarke, A. M., Afolayan, A. J. (2016). Identification of bacterial contaminants in polyherbal medicines used for the treatment of tuberculosis in Amatole District of the Eastern Cape Province, South Africa, using rapid 16S rRNA technique. Journal of Health, Population and Nutrition, 35: 27.

Finnish Medical Agency (FIMEA) (2004). Herbal medicinal products. Article 1.30 of Directive 2004/24/EC. Available at https://www.fimea.fi/web/en/marketing_authorisations/herbal_medicinal_products/herbal_medicinal_products. Visited 28.01.2020.

Firempong, C. K. (2020). Promoting Ghanaian Herbal Medicine (GHM) in the fight against the Covid-19 pandemic. MyJoyOnline.Com. https://www.myjoyonline.com/opinion/promoting-ghanaian-herbal-medicine-ghm-in-the-fight-against-the-covid-19-pandemic/

Gannabathula, S., Skinner, M. A., Rosendale, D., Greenwood, J. M., Mutukumira, A. N., Steinhorn, G., et al. (2012). Arabinogalactan proteins contribute to the immunostimulatory properties of New Zealand honeys. Immunopharmacology and Immunotoxicology, 34, 598–607.

Ghana Environmental Protection Agency. (2008). Ghana environmental protection agency bans 25 dangerous agro-chemicals. Ghana Daily Graphic. Extracted, 27.02.2019.

Govender, S., Du Plessis-Stoman, D., Downing, T. G., Van De Venter, M. (2006). Traditional herbal medicines: Microbial contamination, consumer safety and the need for standards: Research letter. South African Journal of Science, 102: 253–255.

Gyenwali, D., Vaidya, A., Tiwari, S., Khatiwada, P., Lamsal, D. R., Giri, S. (2017). Pesticide poisoning in Chitwan, Nepal: a descriptive epidemiological study. BMC Public Health 17, 619. https://doi.org/10.1186/s12889-017-4542-y

Hakalehto, E. (2015). Microbial presence in foods and in their digestion. In: Hakalehto, E. (ed.). Microbiological Food Hygiene. New York, NY, USA: Nova Science Publishers, Inc.

Hakalehto, E. (2021). Chicken IgY antibodies provide mucosal barrier against SARS-CoV-2 virus and other pathogens. IMAJ, 23, April 2021.

Hakalehto, E., Nyhlom, O., Bonkoungou, I. J. O., Kagambega, A., Rissanen, K., Heitto, A., Barro, N., Haukka, K. (2014). Development of microbilogical field methodology for water and food chain hygiene analysis of Campylobacter spp and Yersinia spp in Burkina Faso, West Africa. Pathopysiology, 21: 219–229.

Heyman-Lindén, L., Kotowska, D., Sand, E., Bjursell, M., Plaza, M., Turner, C., Holm, C., Fåk, F., Berger, K. (2016). Lingonberries alter the gut microbiota and prevent low-grade inflammation

in high-fat diet fed mice. Food and Nutrition Research, 60: 29993. doi: 10.3402/fnr.
v60.29993.

Huang, J., Tao, G., Liu, J., Cai, J., Huang, Z., Chen, J. (2020). Current prevention of COVID-19: Natural
products and herbal medicine. Frontiers Pharmacol, 11.

Huttunen, S., Toivanen, M. C., Arkko, S., Ruponen, M., Tikkanen-Kaukanen, C. (2011). Inhibition
activity of wild berry juice fractions against Streptococcus pneumoniae binding to human
bronchial cells. Phytotherapy Research, 25: 122–127. doi: 10.1002/ptr.3240.

Huttunen, S., Riihinen, K., Kauhanen, J., Tikkanen-Kaukanen, C. (2013). Antimicrobial activity of
different Finnish monofloral honeys against human pathogenic bacteria. Apmis, 121: 827–834.

Huttunen, S., Toivanen, M., Liu, C., Tikkanen-Kaukanen, C. (2016). Anti-infective potential of
salvianolic acid B against Neisseria meningitidis. BMC Research Notes, 9 (25). doi: 10.1186/
s13104-016-1838-4.

Ikeda, T., Ando, J., Miyazono, A., Zhu, X.-H., Tsumagari, H., Nohara, T., Yokomizo, K.,
Uyeda, M. (2000). Anti-herpes virus activity of solanum steroidal glycosides. Biological and
Pharmaceutical Bulletin, 23: 363–364.

Kalumbi, M. H., Likongwe, M. C., Mponda, J., Zimba, B. L., Phiri, O., Lipenga, T., Mguntha, T.,
Kumphanda, J. (2020). Bacterial and heavy metal contamination in selected commonly sold
herbal medicine in Blantyre, Malawi. MMJ, 32: 153–159. doi: 10.4314/mmj.v32i3.8.

Kapil, A. (2005). The challenge of antibiotic resistance; need to contemplate. The Indian Journal of
Medical Research, 121: 83–91.

Katerere, D., Stockenström, S., Thembo, K., Rheeder, J., Shephard, G., Vismer, H. (2008).
A preliminary survey of mycological and fumonisin and aflatoxin contamination of African
traditional herbal medicines sold in South Africa. Human & Experimental Toxicology,
27: 793–798.

Khan, F., Hill, J., Kaehler, S., Allsopp, M., Van Vuuren, S. (2014). Antimicrobial properties and
isotope investigations of South African honey. Journal of Applied Microbiology, 117: 366–379.
doi: 10.1111/jam.12533.

Koffuor, G. A., Dickson, R., Gbedema, S. Y., Ekuadzi, E., Dapaah, G., Otoo, L. F. (2014). The
Immunostimulatory and antimicrobial property of two herbal decoctions used in the
management of HIV/AIDS in Ghana. African Journal of Traditional, Complementary and
Alternative Medicines, 11: 166–172.

Komulainen, P. (2017, October 20). Help sought from complementary and alternative medicine to
remedy health problems. University of Helsinki. https://www2.helsinki.fi/en/news/society-
economy/help-sought-from-complementary-and-alternative-medicine-to-remedy-health-
problems

Kosalec, I., Cvek, J., Tomić, S. (2009). Contaminants of medicinal herbs and herbal products. Arhiv
Za Higijenu Rada I Toksikologiju, 60: 485–501.

Kwakman, P. H., Te Velde, A. A., De Boer, L., Vandenbroucke-Grauls, C. M., Zaat, S. A. (2011).
Two major medicinal honeys have different mechanisms of bactericidal activity. PLoS One,
6: e17709. doi: 10.1371/journal.pone.0017709.

Lantto T. A. (2017). Cytotoxic and apotoxic effects of selected phenolic compounds and extracts
from edible plants. Doctoral thesis, University of Helsinki, Helsinki, Finland.

Mahady, G. B. (2001). Global Harmonization of Herbal Health Claims. Nutrition Journal,
131: 1120S–1123S.

Majtan, J., Kovacova, E., Bilikova, K., Simuth, J. (2006). The immunostimulatory effect of the
recombinant apalbumin 1-major honeybee royal jelly protein-on TNFα release. International
Immunopharmacology, 6: 269–278.

Majtan, J., Bohova, J., Garcia-Villalba, R., Tomas-Barberan, F. A., Madakova, Z., Majtan, T., et al.
(2013). Fir honeydew honey flavonoids inhibit TNF-α-induced MMP-9 expression in human

keratinocytes: A new action of honey in wound healing. Archives of Dermatological Research, 305, 619–627.

Mandal, M. D., Mandal, S. (2011). Honey: Its medicinal property and antibacterial activity. Asian Pacific Journal of Tropical Biomedicine, 1: 154–160. doi: 10.1016/S2221-1691(11)60016-6.

Matsunaga, A., Haruyama, T., Kobayashi, N. (2014). Anti-influenza viral effects of honey in vitro: Potent high activity of manuka honey. Archives of Medical Research, 45: 359–365. doi: 10.1016/j.arcmed.2014.05.006. Erratum in: Arch Med Res. (2014) 45516.

Matziouridou, C., Marungruang, N., Nguyen, T. D., Nyman, M., Fåk, F. (2016). Lingonberries reduce atherosclerosis in Apoe(-/-) mice in association with altered gut microbiota composition and improved lipid profile. Molecular Nutrition and Food Research, 60: 1150–1160.

Mesaik, M. A., Dastagir, N., Uddin, N., Rehman, K., Azim, M. K. (2015). Characterization of immunomodulatory activities of honey glycoproteins and glycopeptides. Journal of Agricultural and Food Chemistry, 63: 177–184. doi: 10.1021/jf505131p.

Mgbeahuruike E.E, Fyhrquist P, Julkunen-Tiitto R, Vuorela H, and Holm Y. (2018). Alkaloid-rich crude extracts, fractions and piperamide alkaloids of Piper guineense possess promising antibacterial effects. Antibiotics, 7(4), 98.

Mgbeahuruike, E. E., Stålnacke, M., Vuorela, H., Holm, Y. (2019a). Antimicrobial and synergistic effects of commercial piperine and piperlongumine in combination with conventional antimicrobials. Antibiotics (Basel), 8: 55. doi: 10.3390/antibiotics8020055.

Mgbeahuruike, E. E., Holm, Y., Vuorela, H., Amandikwa, C., Fyhrquist, P. (2019b). An ethnobotanical survey and antifungal activity of Piper guineense used for the treatment of fungal infections in West-African traditional medicine. Journal of Ethnopharmacology, 229,157-166. doi: 10.1016/j.jep.2018.10.005.

Mokaya, H. O., Bargul, J. L., Irungu, J. W., Lattorff, H. M. G. (2020). Bioactive constituents, in vitro radical scavenging and antibacterial activities of selected Apis mellifera honey from Kenya. International Journal of Food Science & Technology, 55: 1246–1254. doi: 10.1111/ijfs.14403.

Mosihuzzaman, M., Choudhary, M. I. (2008). Protocols on safety, efficacy, standardization, and documentation of herbal medicine (IUPAC Technical Report). Pure and Applied Chemistry, 80: 2195–2230. doi: 10.1351/pac200880102195.

Nweze, J. A., Okafor, J. I., Nweze, E. I., Nweze, J. E. (2016). Comparison of antimicrobial potential of honey samples from apis mellifera and two stingless bees from Nsukka, Nigeria. Journal of Pharmacognosy and Natural Products, 2: 124. doi: 10.4172/2472-0992.1000124.

Obey, J. K., Von Wright, A., Orjala, J., Kauhanen, J., Tikkanen-Kaukanen, C. (2016). Antimicrobial activity of Croton macrostachyus extracts against several human pathogenic bacteria. Journal of Pathogens, 1453428. doi: 10.1155/2016/1453428.

Obey, J. K., Ngeiywa, M. M., Kiprono, P., Omar, S., Von Wright, A., Kauhanen, J., Tikkanen-Kaukanen, C. (2018). Antimalarial activity of Croton macrostachyus stem bark extracts against Plasmodium berghei in vivo. Journal of Pathogens, 2393854. doi: 10.1155/2018/2393854.

Oinaala, D., Lyhs, U., Lehesvaara, M., Tikkanen-Kaukanen, C. (2015). Antimicrobial activity of different organic honeys against Clostridium perfringens. Organic Agriculture, 5 (153–159). doi: 10.1007/s13165-015-0103-9.

Onyemelukwe, N. F., Chijioke, O. U., Dozie-Nwakile, O., Ogboi, S. J. (2019). Microbiological, parasitological and lead contamination of herbal medicines consumed in Enugu, Nigeria. Biomedical Research (Tokyo, Japan), 30: 828–833.

Osuide, G. E. (2002). Chapter 21 – Regulation of herbal medicines in Nigeria: The role of the National Agency for Food and Drug Administration and Control (NAFDAC). In: Iwu, M. M., Wootton, J. C. (Eds.). Advances Phytomed. Vol. 1, 249–258.

Ota, M., Ishiuchi, K., Xu, X., Minami, M., Nagachi, Y., Yagi-Utsumi, M., Tabuchi, Y. et al. (2019). The immunostimulatory effects and chemical characteristics of heated honey. Journal of Ethnopharmacology, 228: 11–17. doi: 10.1016/j.jep.2018.09.019.

Parida, M. M., Upadhyay, C., Pandya, G., Jana, A. M. (2002). Inhibitory potential of Neem (azadirachta indica juss) leaves on Dengue virus type-2 replication. Journal of Ethnopharmacology, 79: 273–278.

Riihinen, K., Ryynänen, A., Toivanen, M., Könönen, E., Törrönen, R., Tikkanen-Kaukanen, C. (2011). Antiaggregation potential of berry fractions against pairs of Streptococcus mutans with Fusobacterium nucleatum or Actinomyces naeslundii. Phytotherapy Research, 25: 81–87. doi: 10.1002/ptr.3228.

Runyoro, D. K. B., Matee, M. I. N., Ngassapa, O. D., Joseph, C. C., Mbwambo, Z. H. (2006). Screening of Tanzanian medicinal plants for anti-Candida activity. BMC Comple Altern Med, 6.

Salonen, A., Virjamo, V., Tammela, P., Fauch, L., Julkunen-Tiitto, R. (2017). Screening bioactivity and bioactive constituents of Nordic unifloral honeys. Food chemistry, 237: 214–224. doi: 10.1016/j.foodchem.2017.05.085.

Samuelsson, G., Bohlin, L. (2009). Drugs of Natural Origin. A Treatise of Pharmacognosy, Apotekarsocieten. Swedish Pharmaceutical Society, Swedish Pharmaceutical Press, 6th, edition.

Schoepfer, A. M., Engel, A., Fattinger, K., Marbet, U. A., Criblez, D., Reichen, J., Zimmermann, A., Oneta, C. M. (2007). Herbal does not mean innocuous: Ten cases of severe hepatotoxicity associated with dietary supplements from Herbalife products. Journal of Hepatology, 47: 521–526. doi: 10.1016/j.jhep.2007.06.014.

Schwede, S., Thorin, E., Lindmark, J., Klintenberg, P., Jääskeläinen, A., Suhonen, A., Laatikainen, R., Hakalehto, E. (2017). Using slaughterhouse waste in a biochemical based biorefinery - results from pilot scale tests. Environmental Technology, 38: 1275–1284.

Shahidi Bonjar, G. H. (2004). Evaluation of antibacterial properties of Iranian medicinal plants against Micrococcus luteus, Serratia marcescens, Klebsiella pneumoniae and Bordetella bronchoseptica. Asian Journal of Plant Sciences, 3: 82–86.

Siddiqui, N. A., Al-Yousef, H. M., Alhowiriny, T. A., Alam, P., Hassan, W. H. B., Amina, M., Hussain, A., Abdelaziz, S., Abdallah, R. H. (2018). Concurrent analysis of bioactive triterpenes oleanolic acid and β-amyrin in antioxidant active fractions of Hibiscus calyphyllus, Hibiscus deflersii and Hibiscus micranthus grown in Saudi Arabia by applying validated HPTLC method. The Saudi Pharmaceutical Journal, 26: 266–273.

Skovgaard, M., Renjel Encinas, S., Jensen, O. C., Andersen, J. H., Condarco, G., Jørs, E. (2017). Pesticide residues in commercial lettuce, onion, and potato samples from Bolivia – A threat to public health? Environmental Health Insights, 11. doi: 10.1177/1178630217704194.

Street, R. A., Stirk, W. A., Van Staden, J. (2008). South African traditional medicinal plant trade – Challenges in regulating quality, safety and efficacy. Journal of Ethnopharmacology, 119: 705–710.

Tchounwou, P. B., Yedjou, C. G., Patlolla, A. K., Sutton, D. J. (2012). Heavy Metals toxicity and the environment. EXS, 101: 133–164. doi: 10.1007/978-3-7643-8340-4_6.

Tilburt, J. C., Kaptchuk, T. J. (2008). Herbal medicine research and global health: An ethical analysis. Bulletin of the World Health Organization, 86: 594–599.

Toivanen, M., Huttunen, S., Lapinjoki, S., Tikkanen-Kaukanen, C. (2011). Inhibition of adhesion of Neisseria meningitidis to human epithelial cells by berry juice polyphenolic fractions. Phytotherapy Research, 25: 828–832. doi: 10.1002/ptr.3349.

Tsimbiri, P.F., Moturi, W.N., Sawe, J., Henley, P. and Bend, J.R. (2015). Health Impact of Pesticides on Residents and Horticultural Workers in the Lake Naivasha Region, Kenya. Occup. Dis. Environ. Med. 3, 24–34.

Turkson, B. K., Mensah, M. L. K., Sam, G. H., Mensah, A. Y., Amponsah, I. K., Ekuadzi, E., Komlaga, G., Achaab, E. (2020). Evaluation of the microbial load and heavy metal content of two polyherbal antimalarial products on the ghanaian market. In: Evid-Based Complement. Altern. Med. 2020, e1014273.

United Nations (2001). Stockholm convention on persistent organic pollutants. Available at https://www.unido.org/our-focus-safeguarding-environment-implementation-multilateral-environmental-agreements/stockholm-convention. Last visited 02.12.2019.

Van Vuuren, S., Williams, V. L., Sooka, A., Burger, A., Van Der Haar, L. (2014). Microbial contamination of traditional medicinal plants sold at the Faraday muthi market, Johannesburg, South Africa. South African Journal of Botany, 94: 95–100.

Verma, V. (2020). Plant therapy of corona virus. International Journal of Pharmaceutical Sciences Research, 10. doi: 10.31531/2231-5896.1000102.

Wagate, C. G., Mbaria, J. M., Gakuya, D. W., et al. (2010). Screening of some Kenyan medicinal plants for antibacterial activity. Phytotherapy Research, 24, 150–153.

World Health Organization (2002). WHO Programme on Traditional Medicine, traditional medicine strategy 2002–2005. https://apps.who.int/iris/handle/10665/67163

World Health Organization (2004). WHO Library Cataloguing-in-Publication Data Joint FAO/WHO Expert Consultation on Human Vitamin and Mineral Requirements (1998): Bangkok, Thailand.

World Health Organization. (2007). WHO guidelines for assessing quality of herbal medicines with reference to contaminants and residues. https://apps.who.int/iris/handle/10665/43510. Extracted 30.03.2021.

Wu, L., Du, B., Vander Heyden, Y., Chen, L., Zhao, L., Wang, M., Xue, X. (2017). Recent advancements in detecting sugar-based adulterants in honey – A challenge. TrAC – Trends in Analytical Chemistry, 86: 25–38.

Xu, J., Song, X., Yin, Z. Q., Cheng, A. C., Jia, R. Y., Deng, Y. X., Ye, K. C., Shi, C. F., Lv, C., Zhang, W. (2012). Antiviral activity and mode of action of extracts from neem seed kernel against duck plague virus in vitro1. Poultry Science, 91: 2802–2807.

Yahfoufi, N., Alsadi, N., Jambi, M., Matar, C. (2018). The immunomodulatory and anti-inflammatory role of polyphenols. Nutrients, 10 (11): 1618. doi: 10.3390/nu10111618.

Yesuf, A., Wondimeneh, Y., Gebrecherkos, T., Moges, F. (2016). Occurrence of potential bacterial pathogens and their antimicrobial susceptibility patterns isolated from herbal medicinal products sold in different markets of Gondar Town, Northwest Ethiopia. International Journal of Bacteriology, 1959418.

Zábrodská, B., Vorlová, L. (2015). Adulteration of honey and available methods for detection – a review. Acta Veterinaria Brno, 83: 85–102.

Elias Hakalehto, Frank Adusei-Mensah, Anneli Heitto,
Ari Jääskeläinen, Jukka Kivelä, Jan Den Boer, Emilia Den Boer

4 Fermented foods and novel or upgraded raw materials for food commodities by microbial communities

Abstract: Microbial presence in any closed or semi-closed system usually strives for the energetically most feasible outcome. In mixed fermentations, this result is achieved by the metabolic networks of versatile microflora. Such biochemical constitution can be used effectively as a lead for food industry development projects, food preservation, and nutritional improvement. Traditionally, microbial inocula have been used for producing fermented foods and drinks. This has often taken place in non-aseptic or semi-aseptic environments. The metabolic network of microbial communities is extremely adaptable with respect to the slightest changes in environmental conditions. In fact, the microbial networks constitute the foundation for the ecological balances in the environment. Moreover, they decisively impact the functions of our alimentary tract, and consequently, the entire body system, too. Actually, microbial ecosystems exist in the environment, in agricultural production, in food, inside our body system, and in the wastes or side streams. They are all important for our food chain.

The basic nature of microbial networks has also been tested in numerous microbial bioprocesses based on the utilization of industrial side streams by the undefined mixed cultures (UMC) approach. Such fermentation processes are used for producing food components. These processes also relate to the environmental or ecosystem engineering processes. We can exploit the principles in sustainable manufacturing industries, which use biomass side streams as raw materials. Then, in the process control, we steer and adjust the functions of the available or boosted microbial ecosystem, which consists of the natural members of the side stream microbiome, as well as the inoculated industrial strains based on our technical decision or choice. The biorefineries offer a continuum for the traditional fermentation technologies in

Elias Hakalehto, Finnoflag Oy, Kuopio and Siilinjärvi, Finland; Department of Agricultural Sciences, University of Helsinki, Helsinki, Finland; University of Eastern Finland, Kuopio, Finland
Frank Adusei-Mensah, Finnoflag Oy, Kuopio and Siilinjärvi, Finland; University of Eastern Finland, Kuopio, Finland
Anneli Heitto, Finnoflag Oy, Kuopio and Siilinjärvi, Finland
Ari Jääskeläinen, Savonia University of Applied Sciences, Kuopio, Finland
Jukka Kivelä, Department of Agricultural Sciences, University of Helsinki, Helsinki, Finland
Jan Den Boer, Department of Applied Bioeconomy, Wrocław University of Environmental and Life Sciences, Poland
Emilia Den Boer, Faculty of Environmental Engineering, Wrocław University of Science and Technology, Poland

https://doi.org/10.1515/9783110724967-005

food production. The products we get, besides the fermented foods or beverages, are also chemical compounds, energy gases, and organic fertilizers. Future biorefineries are industrial fields, where side streams from agriculture, industries, and municipalities are treated and refined into useful products. In a successful project, no waste is left behind. It is all turned into new raw materials. In food manufacturing technologies, the hygienic levels of the substances are also maintained. And the production strategies transfer into sustainable ones with the aid of natural microbial strains.

4.1 Introduction to ecosystem engineering

At the bottom of lake Näsijärvi in Tampere, Finland, about 1.5 million or more tons of the so-called zero fiber have been accumulated during a century as a result of the outlets of a wood industry complex ashore. These residual cellulosic fractions have been preserved in the dark, cold, anoxic, and acidic conditions at the bottom of the lake. This resource and its potential utilization give an eloquent example of the total conversion of industrial wastes or side streams into useful products. Actually, this approach is not only complementary to the petrochemical industries but also illustrates the principle of producing food materials from organic residues. Such an approach could be a kind of extension of traditional fermented foods in the food industry, the latter having been manufactured worldwide throughout the centuries.

In the future, our industries are increasingly relying on both ecosystem engineering and on recycling organic substances. In practice, this means seeing the various side streams as valuable raw materials. Moreover, the harmful substances can be removed by using micro-organisms for the clean-up. Most importantly, in the production of various chemical commodities, we will be using biomass raw materials as complementary sources alongside the fossil deposits or the fresh food substrates. The fossil compounds have to be dug from underground or the sea bottom to be used not only as the sources of food chemicals but also in the preparation of packages for food transport and storage. These materials could be replaced by bio-based ones. For example, for the industrial expanded polystyrene packaging, more eco-friendly replacements are needed. The polylactide foams are one potential solution. In Finland, the State Research Centre (VTT) is developing a solution based on wood pulp. It is also important to keep in mind that by their metabolic versatility, micro-organisms are producing several new materials and intermediary substances useful for the manufacturing of biopolymers and composites.

Correspondingly with the current trend, more and more raw materials and energy are obtained from the organic side streams of agriculture and forestry, fisheries, different manufacturing industries, and so on. In principle, all materials are increasingly seen as resources instead of wastes or useless residues. This actually comes closer to the way, how Mother Nature is functioning. In the ecosystems, most of the organic matter is in circulation by the energy flow originally deriving from the solar energies. And human economy resembles more and more the economy of Nature. At least it should resemble.

Various novel technologies have been developed for implementing the change into the ecosystem engineering kind of approach, which repairs the potential damages caused by Mankind to the ecosystems. The kickoff incident of the new field was undoubtedly the oil accident in the Mexican Gulf on the "Deepwater Horizon" oil drilling platform in 2010 (Hakalehto and Dahlquist, 2018; Passow and Overton, 2021). Then the damages to the ecosystem were believed to be causing devastating long-term effects or even final destructive consequences for the entire Caribbean Sea. However, this ecosystem, in addition to the economic blow of the disaster, opened the eyes of the industrialists to see the possibilities in microbial biotechnology, as the seawater and sediment micro-organisms attenuated the worst consequences of the environmental catastrophe. This underlined the unity of human societies with the natural ecosystems. It also took up most clearly the economic potential in using microbes for cleaning up environmental pollution and eradicating recalcitrant substances (Hakalehto and Dahlquist, 2018). This strategy would benefit all in a long term. For example, the tourism industry, fisheries, as well as all other livelihoods in and around the sea are dependent on the ocean ecosystem. They would get the advantage of this modern thinking and the more developed approach. Environmental health aspects should also be considered and calculated into that equation. Most eloquently, we could turn environmental problems into assets by harnessing the micro-organisms into processes as biocatalysts. They could be applied for producing valuable chemicals, energy in the form of gases such as methane and hydrogen, or by exploiting fatty acids, or microbiologically returning residual biomasses into the production of food or its packages or taken back to the agricultural use as soil improvement or organic fertilizers.

4.2 Utilization of the residual fractions of the biorefinery units as soil improvement agents or organic fertilizers

There is an increasing concern about the dependence of agricultural food production on mineral fertilizers (especially Nitrogen and Phosphorus) and other agrochemicals because these inputs (a) are associated with significant negative environmental impacts; (b) reduce the sustainability of crop production systems, and (c) negatively affect future food security. Therefore, the main concept or strategy available to replace or reduce mineral fertilizer use is to recycle a larger proportion of nutrients that are removed from soils as crops and livestock products back into agricultural lands. We could also take advantage of the microbial Nitrogen fixing in an unprecedented manner (Figure 4.1). This should be based on the efficient processing and recycling of agricultural side streams (e.g., excess manure, residues from on-farm bioenergy generation systems), food processing residues, and domestic or communal waste (including sewage). However, based on current predictions for (a) climate change, (b) increases in world population, and (c) demand for "renewable sources of energy," it will be essential to maintain and improve yields for staple food, feed, and potential biofuel crops. Fertilization regimes based on recycled organic waste should therefore minimize environmental impacts (especially with respect to nutrient pollution and greenhouse gas emissions) while increasing crop yields. The main aim of the proposed project concept is to improve the sustainability of crop production by developing technologies and strategies for the efficient recycling of agricultural and domestic organic wastes as added value organic fertilizers and soil improvers into agriculture.

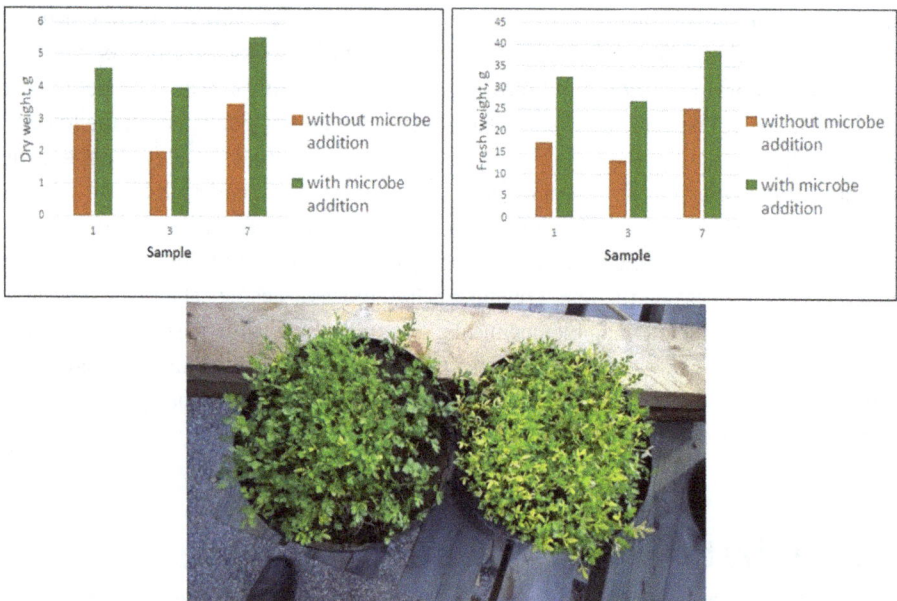

Figure 4.1: Garden grass experiments conducted since September 2015 in the Helsinki University greenhouses at Viikki Campus are being designed and conducted by Jukka Kivelä and Elias Hakalehto. The graphics on the left indicate the plant dry weight in three experiments without microbial addition (orange bars) and with microbial addition (green bars). The right-hand side shows the corresponding results of fresh weight measurements. The lower picture shows garden grass with microbe addition on the left. The nitrogen fixing microbial enforcement to the soil is launched with the trade name Aurobion®. Similar results were obtained with several other plant species, which implies to the potential of micro-organisms in agriculture and plant nutrition. Photo: Elias Hakalehto.

More recycling of food industry byproducts helps in the quantification of the agronomic and the climate change mitigation potential of food production byproducts based on organic fertilizers. This approach will and should focus on crop health, its yield, and quality, as well as selected soil fertility parameters (e.g., soil C, N, and P levels). Use of novel byproduct-based organic fertilizers, have to be quantified and demonstrated. Its effects under different agronomic and pedo-climatic background conditions need to get elucidated. This will focus on detailed assessments of (a) carbon footprints and greenhouse gas mitigation potential; (b) soil chemical, physical, and biological properties; and (c) crop yield and quality parameters.

Existing byproduct fertilizers have shown the potential for reducing mineral fertilizer and crop protection inputs. Especially, under the controlled studies and field trials, the effects of byproduct fertilizers provide a mechanistic understanding of the effects of fertilizers and organic soil amendments which can be exploited to further improve organic waste processing technologies and protocols. To enable the development of rural and urban waste management in different regions based on realistic business plans, it is also proposed to develop a safety and quality assurance system for the production and use of byproduct-based organic fertilizers. This will focus on assuring not only microbiological and toxicological safety parameters but also important quality parameters such as fertilizer value and nutrient release characteristics, suppressiveness, and carbon sequestration potential associated with different fertilizer products. This will in long term make agricultural production more effective, ecological, and sustainable. Moreover, it should be taken into account that some parts of the side streams or waste could be circulated in the areas, for the production of city-food, for instance (Proksch and Baganz, 2020). The vertical farming and cultivation of plants and microbes in cultures and as products by bioprosesing, will open avenues for future food production.

At the same time, it is needed to develop an efficient program of dissemination and training activities to improve the implementation of project results by farmers, rural and urban industries, as well as communities generating organic waste. The waste processing industry needs to be established and developed. The entire food production system requires a clear nutrient recycling program, which allows the use of byproducts as recycled fertilizers. For more discussion on organic fertilization and the role of microbes, see Kivelä and Hakalehto (2016).

4.3 Case Hiedanranta as a valuable example and option of novel thinking in biomass conversion

In the city of Tampere in Southern Finland, the forest industry has accumulated the previously useless "zero fiber" into the lake bottom for almost a century. These cellulosic byproducts have been sedimented in millions of tons (1.5 million tons just in the proximity of the former abandoned Lielahti forest industry complex) onto the lake bottom. During 2018–19, Finnoflag Oy was the key technology provider in the governmentally funded project "Zero waste from zero fiber." This project belonged to the Finnish Ministry of Agriculture "Blue bioeconomy" projects. In the project consortium of companies and research institutes, we developed a three-pilot system for producing food-grade chemicals, energy gases, and organic fertilizers out of the cellulosic waste sediments recovered from the lake bottom:

1. valuable chemicals, such as lactate and mannitol, which both are important raw materials and additives for various foods being non-toxic and versatile in use,
2. energy gases, such as methane biogas, biohydrogen, and hytane (combination of methane and hydrogen), and
3. organic fertilizers or humus-containing high-value products that could be used for returning the soil balance of microbes and organic substances, which is being lost by chemicalization or erosion in spoiled agricultural lands all over the world. This brings the nutrients back to soil, and for producing safe and cleaner plant crops, for instance. Also, the so-called city-food could be produced in the urban areas.

See also the schematic presentation of the project entity (Figure 4.2). The future political decision-making on a larger scale project based on the pilot results was underway when this chapter was written in June 2021. More information on the Hiedanranta biorefinery project could be found in Hakalehto (2018a) or in Beckinghausen et al. (2019). This concept is a case where no waste is remaining or left behind after the environmental deposits of past industrial actions will be utilized in economically feasible ways. These processes are based on research regarding the microbiological communities and processes thereof (Figure 4.3).

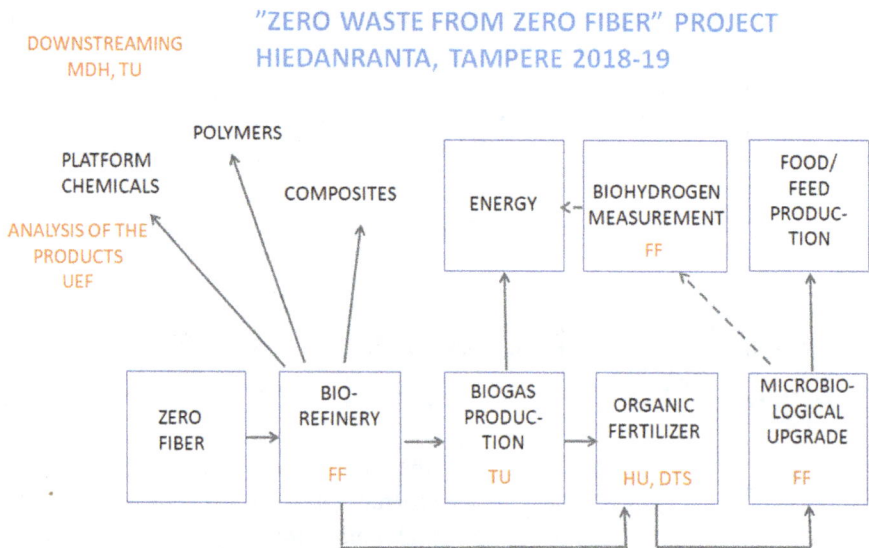

Figure 4.2: "Zero waste from zero fiber"-project was based on the order of the City of Tampere. It was in part funded by the Government of Finland (Ministry of Agriculture and Forestry, ELY Centre "Blue Bioeconomy" project). Ramboll Finland Oyj took care of project coordination. TU = Tampere University; HU = University of Helsinki; UEF = University of Eastern Finland, Kuopio, Finland; MDH = Mälaren University, Västerås, Sweden; DTS = DTS Biosystems Oy; FF = Finnoflag Oy.

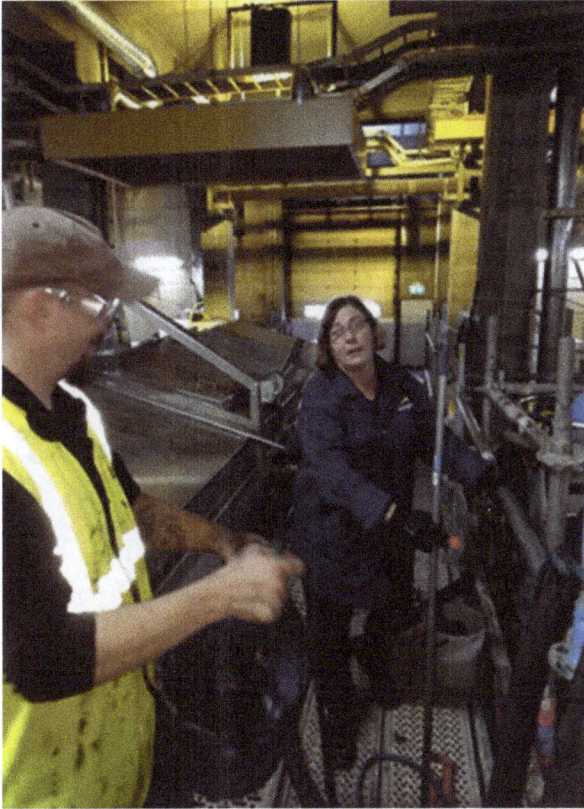

Figure 4.3: The zero fiber deposits from the lake bottom in Hiedanranta, Tampere, were processed in a bioreactor with approximately 15 m³ of effective volume. The exhaust gases were collected and measured. Maximal hydrogen concentration was about 4% in the emitted gas flow (containing the carrier gas). The lactate production rate was 9.2% (w/v). Photo: Elias Hakalehto.

As a side-benefit of these novel industries, the City of Tampere could get cleaned the spacious bay area in front of the designed future lakeside suburb for 25,000 inhabitants. We optimistically and positively look forward to a favourable conclusion by the city decision-makers, which could provide the authorization for the consortium to execute comprehensive plans for this revolutionary industrial project of massive scale and goals to become a global showcase of sustainable development. At the same time, cleaning up the suburban area and the adjacent lake could offer better health perspectives for the citizens of the city. For example, the removal of aerosols generating organic deposits is important for respiratory health. By accomplishing this task with sustainable biorefineries, the city would facilitate 100% recycling of this accumulated waste deposit in a climate-friendly fashion.

4.4 Novel microbial bioindustries in accumulating health and providing nutrition for sustainable societies

Based on cases like the Hiedanranta plan in Tampere, Finland, the new generation could establish novel kinds of industrial activities. Using the microbial communities and their metabolism, these processes could be fashionable for the human living side by side along with the natural ecosystems in the abolished industrial sites (Figure 4.4). Such eye-opening activities could give proofs of massive scale solutions to environmental issues globally. This basic principle of removing waste and converting it into e.g. sustainable food substances and other sustainable products could

Figure 4.4: Hiedanranta Bay area, Näsijärvi, Finland. The abolished factory site is at the top left corner. Center of the city of Tampere is situated on the other side of the lake about 2 km away. In this Hiedanranta area, the lakeside and industrial sites will be converted into a suburb for up to 25,000 residents. The major location of zero fiber deposits on the lake bottom is on the upper part of the bay in this aerial photo, as well as in underground deposits ashore. Total amount of naturally preserved biomass in the close proximity of the former factory site is estimated to be 3.0 million cubic meters. Photo: National Land Survey of Finland.

provide a remarkable option for the future societies to increasingly feed their citizens by the circulated chemical or fertilized commodities, and to compensate for any adversarial climate effect of past activities (Hakalehto, 2018b, 2020).

4.5 Microbial safety through modern food production

There are microbes everywhere and in food they usually reside in high numbers. The human epithelia are also colonized by multitudes of versatile microbiota. Fortunately, since the microbes are essential factors in maintaining our healthy living, the microbiota of ours mostly consists of beneficial species. The members of microbial communities often "work in shifts" in our digestion, according to the food raw materials, body functions, and humoral, immunological, and neurological control (Shin et al., 2019). In other words, members of the microbial communities have specific metabolic preferences. As we consider those preferences, we learn to adjust and optimize the bioprocesses based on the utilization of side streams. In fact, the principles of microbial interactions in the intestines could be utilized in industrial processes as well.

In one spoonful of yogurt, there are as many bacterial cells as humans on the planet Earth. This is a microcosm we all encounter every day. These micro-organisms play huge roles in food production and preservation. Different generations have learned to shelter from diseases and maintained good hygiene long before the connection between microbes and health threats or potentials were demonstrated. Many foods are still produced by using microbes. Different fermentation processes are important in food manufacturing protocols and practices. Many additives are produced by the microbes, which then improve the quality and health benefits of food. For example, citric acid that is mainly produced by biotechnological methods is commonly used for flavour and for preserving foods. New possibilities for sustainable production of raw materials are indeed opened through the various microbiological processes. Biotechnological components in our food, such as IgY-antibodies in eggs, give us a chance to improve global health security by functional food components (Hakalehto, 2021b). As it is stated by Hakalehto and Jääskeläinen (2017): "At present, to restructure our way of living during the post-industrial era, it is crucial to put weight on recirculation. This could open up novel technological solutions, which will help in balancing the human impact on the global ecosystem."

4.6 Microbial metabolism can be utilized

4.6.1 Possibilities in the microbial fermentation

The health benefits of acetic acid are known as components in complementary reme-dies all over the world (Ho et al., 2017). Acetic acid is a product of microbial fermenta-tion processes. As a relatively weak acid, it is suitable for cooking, preserving, and seasoning. It can also be used to extract ingredients in a natural way. Acetic acid is secreted out of the bacterial cells that produce it and is easily collected. Its acidity prevents the excessive growth of microbes in the food, and acetic broth is a suitable method to preserve different vegetables, for instance. The proper preservation in-creases shelf life of the food products. Maintaining good qualities and stability is highly important for the productivity of novel items and food-related services, such as catering operations and home deliveries (Sauramäki and Hakalehto, 2015).

Another widely used method to preserve vegetables, dairy, fish, or meat prod-ucts, is lactic acid fermentation. It is potentially performed using mixed natural lac-tic acid bacterial (LAB) culture, and starter cultures are also often used. It is both a very motivating hobby and also an interesting industrial action to ripen sour veg-gies into healthy dishes using this method for such products as the famous "Sauer-kraut" which could make our diet healthier and more versatile (Beganović et al., 2011). Many LAB are extremely beneficial for the specific health and balance in the intestinal tract. It is no coincidence that most probiotics are LAB. Both acetic and lactic acids can be used as such as preservatives. Also, many other organic acids, such as benzoic acid, provide preservative functions. The various traditional fer-mentation methods are presented in Section 4.9.

Propionic acid or propionate, as a preservative used in meat production and bak-ing industries, is also produced by microbes. In the EU's Baltic Sea biorefinery project's (ABOWE, 2014) piloting test runs in Poland or Sweden propionic acid was successfully produced from the potato side stream and the wastes of the poultry industries (chicken litter), combined with domestic or separately collected organic residues of the restau-rant waste (den Boer et al., 2016a, b, 2020; Schwede et al., 2017). Pilot runs were also carried out in Finland using the waste from the forest industries (Hakalehto et al., 2016a). These activities also served the later development of such efforts as the Hiedan-ranta biorefinery project in 2018–19 (see above). The biological preservative substances are often important in maintaining the quality of the distributed food.

Like organic acids, alcohols produced by the microbes can be used for flavours, preservatives, and as functional ingredients or raw materials. For example, 2,3-butanediol is produced by the metabolism of groups of bacteria such as *Klebsiella/Enterobacter*. They are important in balancing microbiological agents for the human intestinal tract, too. The secreted 2,3-butanediol compensates the stress caused by the acid formation on the bacteriological intestinal balance (BIB) (Hakalehto et al., 2008, 2010, 2013; Hakalehto 2012). One of the future challenges of the food industry is to understand, how different food ingredients influence the human body system. For example, ethanol which is formed as a side product of 2,3-butanediol fermentation, could cause obesity as it contains lots of chemical energy. In turn, lactic acid has the smallest density of energy of the organic acids produced by microbes in the human body. Mannitol is an example of a low-calorie sweetener (Mooradian et al., 2017).

We produced 2,3-butanediol from potato waste in the laboratory of Finnoflag Oy in 2008 at a world record level (8 g/L/h) (Hakalehto et al., 2013). This platform chemical has high value as an intermediate product and as a multi-functional chemical in cosmetic industry, anti-ice agent, synthetic rubber raw material as a source material for butadiene, and for polymer industries. 2,3-Butanediol is also called 2,3-butylene glycol. It is a common non-toxic substance found in skin creams, beauty products, and similar products (Figure 4.5).

Figure 4.5: After easening of the coronavirus precautions, it was possible to travel again to the Taffel Ab's potato chip factory's playground on Åland Island in southwestern Finland. Already in 2008, at Finnoflag Oy's lab, we achieved a world record productivity of 2,3-butanediol (8 g/L) out of another potato industry waste stream in Finland (Hakalehto et al., 2013). Similar type of result was obtained from chips factory waste during Implementing Advanced Concepts for Biological Utilization of Waste (ABOWE) project trials in Poland (den Boer et al., 2016). Photo: Jukka-Pekka Hakalehto.

In the future, it is possible to design, not only the food ingredients, but also the bacteria integrated with them, based on their health and safety effects (Hakalehto, 2020). The future probiotics and prebiotics could contain positive health effects, as well as long-term influences, such as anti-tumorigenicity (Khalifa et al., 2019; Villeger et al., 2019; Hakalehto et al., 2018). Moreover, such a concept as the panbiotics has been introduced by Hakalehto (2021a). This refers to beneficial or probiotic strains commonly or even universally distributed in the population.

4.6.2 Biological components and microbial balance of the food

In the human digestive system, microbes entering the intestinal tract encounter the local microbiome that has already established itself in the alimentary tract. The microbes in the respiratory tract have a high immunological significance also for digestive functions and overall health (Sansonetti, 2019). Those include toxin-producing species of such bacterial genera as *Streptococcus* and *Staphylococcus* (Spaan et al., 2017). Balanced microbiomes such as the ones under the concept of the BIB help the body to prevent problems caused by the toxin-producing pathogens (Hakalehto, 2012, Hakalehto 2021a, b). The right kind of nutrition maintains the balance of the microbiome. This balance is often broken because of a bad diet causing different intestinal inflammation or dysbiosis (Hoarau, 2016). This creates a "malicious balance" of dysbiotic microbes (Schippa and Conte, 2014). The original, health-beneficial microbial diversity and functionality should then be restored by dietary means, and versatile treatments, and probiotics. In the 2013 European Society for Clinical Nutrition and Metabolism (ESPEN) meeting in Leipzig, Germany, we introduced a method that could quantitate the advantageous effect of the probiotics *in situ*, by a simulation of the activities in the intestinal normal flora. The method was based on the use of portable microbe enrichment unit (PMEU) devices with the capability to simulate the microbial colonies (Hakalehto and Jaakkola, 2013).

4.7 Hiedanranta biorefinery plant, a prodigy of recycling projects

Future food production will include collecting and recycling waste from households, industry, and agriculture and using these residues as raw materials (Hakalehto and Jääskeläinen, 2017; Hakalehto, 2020). This is a process that was successfully studied in the ABOWE project in the years 2012–14 (see Section 4.11). Results from this European Union biorefinery project were later used on making ground for practical procedures in Wroclaw, Poland, for instance (den Boer and den Boer, 2018). These side streams do not have to be food-based. For instance, food-additives free from toxic effects (lactic acid, mannitol) were produced in Hiedanranta. In Tampere in the project "Zero waste from zero fiber" funded by the Finnish Ministry of Forestry and Agriculture (Hakalehto, 2018a, Beckinghausen et al., 2019), the aim was to study the conversion of millions of tons of past industrial deposits of the lake sediment, into useful products (See also Sections 4.3 and 4.4.). The project was a part of the Centre for Economic Development, Transport, and the Environment "Sininen Biotalous" (Blue Bio-based Economy) program. Its main raw material was cellulose-based zero fiber that had accumulated into the lake sediments for approximately one hundred years.

In the biorefinery pilot runs executed by Finnoflag Oy in Hiedanranta, microbes and their enzymes were used as biocatalysts. A corresponding UMC (Undefined Mixed Culture) principle was in use also in both biogas pilot of the University of Tampere and residual fraction processing by DTS Biosystems Oy. This entire project entity gave us a holistic view of a type of recycling in which the waste was fully converted into valuable chemicals and fertilizers and energy (Figure 4.6). The chemicals produced in the biorefinery could ensure the economic feasibility of the entire system and the simultaneous ecosystem engineering operation. New methods developed by Finnoflag Oy are also capable of producing remarkable levels of organic hydrogen ("biohydrogen") and organic fertilizers with enhanced nutritional features from different biomass residues. The latter can be used to restore the fertility of the soil in those areas that suffer from declined agricultural producibility. Ekosovellus Oy, Ecolan Oy, and the University of

Figure 4.6: The inspection of the old factory premises in Hiedanranta, Tampere, Finland, in 2018. The biorefinery pilot reactor was jointly designed by Finnoflag Oy and Nordautomation Oy, and it was transferred to the testing site. This unit reactor was then used in the project "Zero waste from zero fiber." The sustainable and economically feasible processing of the cellulosic side streams by microbes was funded by the Finnish Ministry of Agriculture and Forestry. Photo: Lauri Heitto.

Helsinki were involved with Finnoflag Oy in these tests during this project. We have calculated that in theory the Hiedanranta residues, after bioprocessing, could facilitate the returning of 50 km^2 of spoiled agricultural land back into food production.

Besides recycling, it is important for the city of Tampere to eliminate the environmental and safety hazards, such as thinner ice cover in winter caused by the organic emissions from the lake bottom. Also, the displeasing odor from the organic deposits and unhealthy gas emissions could be avoided. These objectives belong to the large entity of ecosystem engineering methods, as described above. At the same time, it can be demonstrated that environmental protection is economically feasible, as the side streams can be utilized in the chemical industry and food production units. Consequently, the following chain will remain unbreakable: healthy soil–ecological food production–safe food distribution–health and sustainability promoting consumption and living–side streams and waste recycling–nutrients to soil again.

4.8 Molecular mechanisms that can be used to improve the safety and hygienic quality of food

4.8.1 Microbiological preservatives

The food we eat turns into chyme in our intestinal tract. From that chyme in the intestines, different nutrients are taken into our body system after a complicated process of extracting and processing the food by the enzymes, solvents, tracers, etc. made by both the body system and the microbiome. Probiotic microbes can be used to balance the processes (Hakalehto and Jaakkola, 2013). The organic acids mentioned elsewhere in this chapter can be useful preservatives with remarkable health benefits. Additionally, many antioxidants can work as preservatives in different herbal and medical products, as well as in food production (Immonen et al., 2016).

4.8.2 Therapeutic and prophylactic antibodies – passive immunization

During the COVID-19 pandemic, new and faster ways to shelter or eradicate various pathogens were actively investigated. One possibility is the use of IgY-antibodies obtained from the egg yolk of immunized chicken (Hakalehto 2021a, b). Elias Hakalehto presented this option to prevent pandemics already in the 1990s (Hakalehto and Kuronen, 1997, 1998). Some scientists have also suggested adding IgY-antibodies into different foods. Chicken is one of the first domestic animals in the use of humans, and it offers many options for industrial applications (Figure 4.7) (Hakalehto, 2015a). During the last decades, IgY has been used in the prevention of pathogens such as *Pseudomonas aeruginosa* (Nilsson et al., 2008), *Vibrio cholerae* (Taheri et al., 2020), influenza virus (Nguyen et al., 2010), caries initiating *Streptococcus mutans* (Hatta et al., 1997), severe acute respiratory syndrome (SARS)-1 virus (Fu et al., 2006), and many other pathogens.

Figure 4.7: Small henhouses could be important in the future production of eggs with therapeutic and prophylactic potential. Photo: Helena Arffman.

4.8.3 New generation of antibiotics and antimicrobial substances in the food

Since producing new antibiotics is expensive to research and requires demanding development work, there are often not enough resources allocated for this important objective. The spreading of new, antibiotic-resistant bacteria often takes place during and after the pandemics among the population. Finland has long traditions in utilizing the antimicrobial effects of various foods. For example, the acetification (acetic acid formation by aerobic bacteria) of sour milk was prevented by the old folks in Finland by placing a frog into a storage container of sour milk in a cold cellar. The amphibian magainin peptides of the frog's skin effectively prevented the growth of many bacteria, such as salmonellae or Methicillin-resistant *Staphylococcus aureus* (MRSA) (Samgina et al. 2012, 2016). The bacteriological part of the experiment was carried out using the PMEU device in Finnoflag Oy's laboratory in Finland.

Using microbiological knowledge in food production offers huge opportunities to develop safer, more affordable high-quality products. The international food industry is in a great position for this development work which also serves the purposes of sustainable development globally by using the potentials of microbial strains and communities (Hakalehto, 2020, 2021c).

4.9 Fermented traditional foods offer a platform for product development in food industries

Traditionally, foods were preserved through naturally occurring fermentation processes which have been used for thousands of years (Table 4.1). In fact, biological fermentation was one of the earliest technologies of preserving food prior to the invention of artificial refrigeration in the mid-1750s by Scottish Professor William Cullen (*Wikipedia: Refrigerator, 2021*). Presently, fermented foods and beverages are part of our daily human diet globally. They provide important sources of nutrients, economic value, and great potential in maintaining health and in preventing diseases. Many different traditional fermented foods and beverages are produced in households or in industrial plants globally. Fermented milk (yogurt, torba yogurt, kurut, ayran, kefir, koumiss), cereal-based fermented foods (tarhana), non-alcoholic beverages (boza), fermented fruits, and vegetables (turşu, şalgam, hardaliye), and fermented meat (sucuk) are delicacies in different parts of the world. Fermented foods are usually consumed as fresh, boiled, sauces, fried, in stews and soups, or as drinks. These foods are known for their nutritional, health, and economic values but they are also still more important because of the tradition and cultural values they create and maintain for the people.

Despite the universal acceptance, however, there are some differences in the approach and preparation of traditional foods and beverages between regions and countries (Kabak and Dobson, 2011). Different types of fermented foods such as chongkukjang, doenjang, ganjang, gochujang, and kimchi are common in Japan and parts of Asia. Fermented fruits and vegetables like turşu, şalgam, hardaliye are delicacies in Turkey and some parts of Middle Eastern countries. Currently, fermented foods and beverages including "banku," kenkey, fura, pito, koko, wagashi, dawadawa, and gari constitute close to 40% of daily diet by Ghanaian homes (Ansong, 2020). Concerning flavour development of fermented corn dough used in Ghana, aroma components like lactic acid, acetic, butyric and propionic acids have been reported (Halm et al., 1993).

Table 4.1: Overview of traditional fermented foods (Kabak and Dobson, 2011; Guzel-Seydim et al., 2011; Patra et al., 2016).

Traditional food	Country/ region	Process/agent	Nutritional/health benefit
Cereal products			
Braga or brascha	Eastern European countries	Fermentation of millet, cooked cereal by yeast and LAB	Lactic acid, fat, protein, carbohydrate, and fiber
Busa, bouza	Egypt, Balkans Bulgaria	Yeast and LAB fermentation of cereals and cocoa	Lactic acid, fat, protein, carbohydrate, and fiber
Kenkey and koko	Ghana	Fermented maize or millet dough with LAB, obligative heterofermentative lactobacilli, yeast	Proteins, vitamins, and fiber Improves gut health
Tarhana	Turkey	Fermentation of cereal-based food	Rich in vitamins and minerals
Togwa	Tanzania	Sorghum or maize by LAB fermentation Consumed fresh	Antimicrobial, improves intestinal mucosal barrier function
Uji	East Africa	Fermented maize, sorghum or millet by *Lactobacillus plantarum*, *L. paracasei*, *L. fermentum*, *L. buchneri*, *Pediococcus acidilactici*, *P. pentosaceus*	Proteins, minerals, vitamins, and fiber Improves cardiac and gut health
Fermented milk			
Torba yogurt, kurut, ayran, koumiss	Anatolia, Turkey, Middle East	Combined bacterial lactic acid and yeast ethanol fermentation of milk lactose by microbial activity of "kefir" grains	Immune booster, lowering of cholesterol, antimutagenic and anticarcinogenic properties, vitamins, and minerals
Ogi	Nigeria, Benin	Maize, sorghum, or millet fermentation with *P. pentosaceus*, *L. fermentum*, *L. plantarum*, yeast	Vitamins, riboflavin, folic acid, minerals, and essential amino acids
Ben-saalga	Burkina Faso	Fermented millet with *L. fermentum*, *L. plantarum*, and *P. pentosaceus*	Rich proteins and mineral source
Yogurt	Global	Milk cultured with LAB	Protein, immune booster, vitamins, and minerals

Table 4.1 (continued)

Traditional food	Country/ region	Process/agent	Nutritional/health benefit
Cheese	Global	Fermented milk with variety of bacteria or mold or yeasts	Protein, vitamins, minerals, and omega-3 fatty acids
Piimä and viili	Finnish specialty	Mesophilic LAB, e.g., *Lactococcus lactis* subsp. *cremoris*, *L. lactis* biovar. *diacetylactis*	Minerals, proteins, vitamins, improved taste, and shelf life
Amasi	South Africa, Zimbabwe, Kenya	LAB milk fermentation for several days in gourd calabashes or in stone jars	Probiotic Reduces bouts with diarrhea
Filmjölk	Scandinavia	Mesophilic fermentation of milk with *L. lactis* and *Leuconostoc* sp.	Vitamins and improved gut health
Kefir	Originally from Caucasus region	Fermented beverage with Kefir grains, a mixture of bacteria and yeasts	Improved digestion and bone health, anti-inflammatory effect
Kumis	Central Asia	Carbonated horse milk fermented with lactobacilli and yeasts	Gut health, healthy cardiovascular and nervous systems, immune booster
Fermented tubers			
Gari, lafun/ kokonte	West and parts of East Africa	Produced from different lactic acid and yeast fermentation of cassava	Improve shelf life, nutritional value, and flavour
Agbelima	Ghana	Cassava fermented with *Lactobacillus plantarum*, *L. brevis*, *L. fermentum*, *Leuconostoc mesenteroides*, also *Bacillus* spp., *Penicillium* spp.	Increases availability of vitamins and minerals
Fermented meat			
Sucuk	Turkey	Semi-dried beef sausage by dry curing Spiced, macerated meat fermented naturally for several weeks	Rich in protein, minerals, and flavour
Fermented vegetables			
Iru, Ogiri, and Ugha	Nigeria	Fermented seeds with LAB	Rich in minerals, vitamins, and improved eye health

Table 4.1 (continued)

Traditional food	Country/ region	Process/agent	Nutritional/health benefit
Kimchi	Japan	Short fermentation of napa cabbage and other ingredients with variety of bacteria strains	Antioxidant, anti-obesity, anticancer, antibacterial, immune boosting, cholesterol lowering
Sauerkraut	Japan, parts of Europe, and the USA	Shredded cabbage fermented by LAB	Fiber, vitamins, and antioxidants Promote eye health
Fermented soybean products			
Tempe	Indonesia	Fermented soy product with *Klebsiella pneumoniae* subsp. *ozaenae,* and *Enterobacter cloacae*	Vitamins, minerals, and high protein Reduced risk to heart disease.
Natto	Japan	Staple probiotic from fermented soybeans	Fiber and vitamin K Support digestive health and lower blood pressure
Miso	Japan	Fermentation of salted soybeans with koji fungus	Lower risk for breast cancer, stroke, and blood pressure
Fermented fish			
Surströmming	Sweden	Fermentation of lightly salted Baltic Sea herring with LAB	Protein, minerals, and vitamins Stomach health
Fish sauce	Indonesia	Fermentation with variety of microbes	Flavour, tasty, and nutritious
Anchovy fish products	Korean	*Bacillus cereus, Clostridium setiens, Pseudomonas halophillus,* and *Serratia marcescens*	Vitamins, proteins, and minerals
Fermented drinks and beverages			
Sima	Finland	Non-alcoholic sparkling beverage from water, sugar or honey, lemon with yeast fermentation	Rich in healthy probiotics, vitamins, and minerals
Boza/bassoi	Turkey	Non-alcoholic Turkish beverage Yeast and LAB fermentation of cereals	Lactic acid, fat, protein, and carbohydrate
Wine and rice wine	Global	*Pseudomonas, Achromobacter, Flavobacterium,* or *Micrococcus* spp., *Leuconostoc mesenteroides*	Rich in antioxidants, anticancer, good heart health

Table 4.1 (continued)

Traditional food	Country/ region	Process/agent	Nutritional/health benefit
Vinegar	Global	Yeast and *Acetobacter ascendens* fermentation of varying substrates	Reduced cholesterol and antimicrobial effect
Chongkukjang	Japan	Short fermentation of boiled soybean seeds using *B. subtilis* and rice straw	Nutrients, immunostimulant, antimicrobial and anti-inflammatory, antioxidant activity
Doenjang	Japan	Fermentation of boiled soybean seeds using naturally occurring bacteria	Anticancer, antimutagenic, antioxidant, and fibrinolytic activity
Ganjang	Japan Korea and other Asian countries	Sauce from fermented soybeans	Anti-colitis, anticancer, and antihypertensive properties
Gochujang	Japan	Fermented paste of red chili powder	Capsaicin reduces body weight, speed metabolism, burn fat

LAB, lactic acid bacteria.

4.10 Biochemicals produced from fermentation

There are numerous positive aspects related to microbial fermentation for food preparation. During fermentation, certain compounds are produced depending on the metabolic pathways of the fermenting agent, as well as on the substrate. Fermented foods are good sources of organic acids which serve as natural preservatives in the fermentation mass against non-beneficial microbes and increase the shelf life of the resultant product. Preservatives alter the pH, water activity, and titratable acidity level, thereby controlling the growth of potential pathogen or food spoiling microbes (Rhee et al., 2011; Immonen et al., 2015, 2016; Wasieleski, 2018). Fermentation undeniably enhances the nutritional value of foods and beverages and their health benefits. Some traditional foods such as Far Eastern soya sauce is recommended to be produced as "naturally brewed" to achieve health benefits and to avoid disadvantageous side effects. During the process, certain short-chain fatty acids (SCFAs) are produced having demonstrated the beneficial impact on health and body energies. SCFAs are small molecular weight carboxylic acids or fatty acids with carbon chain lengths ranging from two to six (Wasieleski, 2018; Hakalehto, 2020). They are also relatively volatile organic compounds due to their small molecular weights (den Boer et al., 2020).

Products of fermentation (fermentation-derived ingredients) could be used to enhance the shelf life, flavour, and other qualities of non-fermented foods when they are added to the non-fermented foods (Table 4.2). The fermentation-derived ingredients alter the product's chemical characteristics (Wasieleski, 2018). Fermentation-derived ingredients are being used today in multiple food categories including bakery, meats, soups, sauces, and dressings sectors.

Key important microbes for many fermentations are the LAB used in multiple tasks, and acetic acid bacteria, for example, *Acetobacter* sp. (*Acetobacter aceti*, *A. xylinum*, and *A. ascendens*) and yeast for alcohol fermentation. Acetic acid bacteria (AAB) are important in the production of vinegar (acetic acid) (FAO, 1998).

The exhaustion of fossil energy resources, global warming, and growing population has called for alternative renewable energy sources to save the global ecosystem. Thanks to microbes, numerous SCFA fermentation products could salvage the energy crisis and non-renewable sources today. Though it is a work in progress, a lot has been done in the past about the production of energy compounds using fermentation products as building blocks. Some of these products include fermentative succinate production from renewable biomass (Li and Xing, 2015), itaconic acid (Steiger et al., 2013) (Okabe et al., 2009), propionic acid (propionate) (Gonzalez-Garcia et al., 2017), valeric acid (Goldberg and Rokem, 2009), and pathogen control with lactic acid (LA) (Wemmenhove et al., 2016).

Table 4.2: SCFAs and their industrial and health role (Saranraj, 2019; Li and Xing, 2015; Ciriminna et al., 2017; Hefetz and Blum, 1978; Yuille et al., 2018; Song et al., 2019; Saha and Nakamura, 2003).

Sources	Active product	Major role
LAB	LA	Preservative, antimicrobial effect
Yeast, *Acetobacter aceti*, *A. xylinum*, and *A. ascendens*	Acetic acid	Antimicrobial effect
Actinobacillus succinogenes, *Escherichia coli*, *Saccharomyces cerevisiae*, *Anaerobiospirillum succiniciproducens*, *Corynebacterium glutamicum*, *Mannheimia succiniciproducens*, and *Basfia succiniciproducens*	Succinic acid (1,4-butanedioic acid)	Building block for the synthesis of high value-added derivatives and fuel, such as 1,4-butanediol (1,4-BDO), tetrahydrofuran (THF), γ-butyrolactone (GBL), succinimide, and biodegradable polybutylene succinate (PBS)
Aspergillus species, (*A. itaconicus* and *A. terreus*)	Itaconic acid (2-methylidenebutanedioic acid)	Monomer for resins, plastics, paints, and synthetic fibers, rubbers, surfactants, and oil additives

Table 4.2 (continued)

Sources	Active product	Major role
Cytromices (Penicillium) mold and A. niger	Citric acid (2-hydroxy -1,2,3-propanetricar-boxylic acid, $C_6H_8O_7$)	Acidulant, preservative, emulsifier, flavouring, food sequestrant and buffering agent, beverage, pharmaceutical, nutraceutical, and cosmetic products
Propionibacterium acidipropionici, P. shermanii, and P. freudenreichii	Propionic acid (propionate)	Antimicrobial, preservative, or herbicide; cosmetics, plastics, food, and pharmaceutical industries
Ants have high concentration of formic acid in their venom. Formate, the reduced form of formic acid, occurs widely in nature.	Formic acid, (methanoic acid)	Preservative, antibacterial agent in livestock feed; in Europe, it is applied on silage control against etc. E. coli.
Megasphaera massiliensis MRx0029	Valeric acid (pentanoic acid, $C_5H_{10}O_2$)	Perfumes and cosmetics, food additives, plasticizers, pharmaceuticals, and probiotics
Klebsiella oxytoca, K. pneumoniae, and Bacillus sp.	2,3-Butanediol (2,3-BD)	High-quality aviation fuel octane can be produced by 2,3-BD. Antifreeze and monomer to synthesize polymer
Lactobacilli such as Lactobacillus brevis and L. buchneri	Mannitol	Many applications in food, pharmaceuticals, medicine, and chemistry

LA, lactic acid.

4.10.1 Bio-dyes from microbes

Microbial pigments are of great structural and colour diversity (Table 4.3). Pigments from bacteria like *Chromobacterium violaceum*, *Janthinobacterium lividum*, *Chromobacterium lividum*, and *Pseudoalteromonas luteoviolacea* have been used for dyeing textiles with good results. Numerous bacterial and other microbial strains produce beautiful colors in culture media and could be an excellent means for the production of chemical dyeing feedstocks in the textile industries (Figure 4.8). Fabrics of natural origin are comparatively high nutrient sources and contain moisture which in turn provide important potentials for fabric degrading and pathogenic microbes. Natural dyes may have antimicrobial activity that would confer protection of fibers against pathogenic microbes. Natural dyes therefore may serve as excellent dyeing agents for gowns meant for hospital usage.

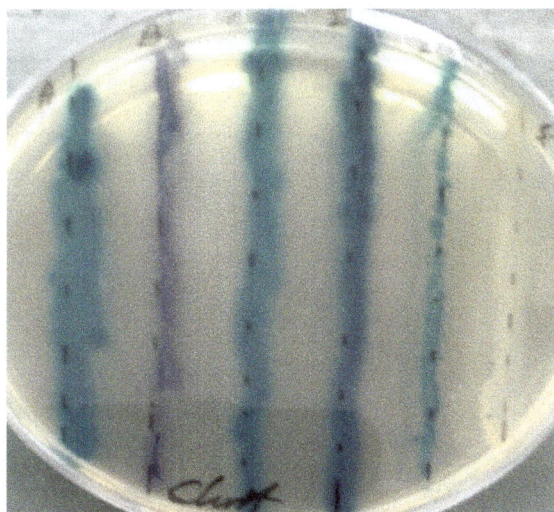

Figure 4.8: Different shades of beautiful colors from strains of LAB on the chromogenic Chromagar™ plate (Becton, Dickinson and Company, Franklin Lakes, NJ, USA). Photo: Frank Adusei-Mensah, at Finnoflag laboratory, Siilinjärvi Finland, 2021.

Until now, most natural renewable dyes are of plant origin, but the current global ecological challenge makes microbial dyeing a powerful alternative source for reconsideration. Using microbes for industrial processes and production offers a more sustainable means of production with minimal investments.

Table 4.3: Naturally derived dyes from micro-organisms (Heer and Sharma, 2017;Usman et al., 2017; Kanelli et al., 2018; Hernández-Velasco et al., 2020).

Micro-organisms	Pigments	Dye color/ appearance
Achromobacter	Zeaxanthin	Creamy
Agrobacterium aurantiacum	Astaxanthin	Pink, red
Bacillus	Zeaxanthin	Brown
Blakesela trispora	Lycopene β-carotene	Red, yellow orange
Bradyrhizobium sp.	Canthaxanthin	Dark red
Brevibacterium sp.	Zeaxanthin	Orange yellow
Chromobacterium violaceum	Violacein	Purple, violet
Corynebacterium michigannise	Zeaxanthin	Grayish to cream

Table 4.3 (continued)

Micro-organisms	Pigments	Dye color/appearance
Corynebacterium insidiosum	Indigoidine	Blue
Dunaliella salina	β-Carotene	Cream
Flavobacterium sp.*Paracoccus zeaxanthinifaciens*	Zeaxanthin	Yellow
Haloferax alexandrinus	Canthaxanthin	Dark red
Janthinobacterium lividum	Violacein	Dark violet, purple
Monascus sp.	Monascorubramin, Rubropunctatin	Yellow, orange, red
Monascus roseus	Canthaxanthin	Orange–pink
Paracoccus carotinifaciens	Astaxanthin	Pink–red
Phaffia rohodozyma,	Astaxanthin	Red
Pseudomonas aeruginosa	Pyocyanin	Blue–green, green
Rhodotorula sp., *R. glutinis*	Torularhodin	Orange–red
Rugamonas rubra, Streptoverticilliumrubrireticuli, Vibrio gaogenes,Alteromonas rubra	Prodigiosin	Red
Serratia marcescens	Prodigiosin	Reddish orange
Staphylococcus aureus	Staphyloxanthin, Zeaxanthin	Golden yellow
Streptomyces coelicolor	Prodiginine, prodigiosin	Pink to red
Xanthophyllomyces dendrorhous	Astaxanthin	Pink–red
Xanthomonas oryzae	Xanthomonadin	Yellow

4.10.2 Bio-preservatives by microbes

Concerning the spoilage organisms, harmful bacteria, and fungi (yeast and mold) are of major concern. Product's innate and external characteristics including temperature, pH, water activity, titratable acidity, are important factors affecting the product spoilage (Immonen et al., 2015). In addition, the spoiling agent's metabolic pathways such as catabolism, anabolism, and overflow metabolism have been reported to influence the ongoing direction of the spoilage reaction and the final state of the spoiled product (Hakalehto, 2015b).

A preservative is a product that can alter the spoilage factors and provide the restrictive environment for non-beneficial spoiling microbes thereby suppressing their growth and prolonging the shelf life of the product (food, pharmaceutical, cosmetic, etc.). Different fermentation products serve as important preservatives for diverse product lines across different industries. Food and product spoilage is undesirable since it leads to unwanted consequences including the destroyal of quality and economic loss for producers. It also causes potential serious health risks and leads to foodborne outbreaks (Hell et al., 2015). The use of beneficial microorganisms to preserve food, drugs, cosmetics, and other products prevents other spoilage organisms from ruining the economic value, taste, and other qualities of the product (Hakalehto et al., 2011). Fungi can produce mycotoxins that can induce serious undesirable health conditions such as foodborne outbreaks (Vanhoutte et al., 2016). During the current pandemics by severe acute respiratory syndrome coronavirus 2 (SARS-CoV-2), the *Mucor* sp. molds have caused severe fungal epidemics in India in the aftermath of the virus variants (Sen et al., 2021). Therefore, it is important to attenuate the excessive growth of fungi in the products. On the other hand, it is possible to exploit fungi to produce new materials (Cerimi et al., 2019).

Previously, preservatives in food, pharmaceutical, and other consumer products have been mostly based on the use of synthetic compounds or plant extracts, antimicrobial substances, and other compounds with their associated strengths and shortfalls. The rising trend of antibiotic resistance, shortage of land for commercial crop cultivation, global warming, and side effects of synthetic compounds has called for continuous search for alternative renewable preservatives and other additives, for example, for the food composition or outlook. Microbes may provide the solution for many issues and reduce the dependency on plants for renewable preservative spices and dye production. Focusing on microbes in this additive area over plants presents a lot of benefits for industries and for the environment. Firstly, small land space is needed for microbial production of preservatives spices and dyes compared to plants. This land is often needed for food production more urgently. Again, microbial fermentations may take only a few hours to days to mature (Figure 4.9), whereas plant cultivation may take from months to many years. Furthermore, depending on microbes, which are naturally in abundance, they will spare land and the plants for Nitrogen fixation, food production, and ultimately reduce the global warming the world is currently facing. For example, in the research by Finnoflag Oy, the autonomous Nitrogen binding by micro-organisms turned out to be highly effective and increased the yields of Chinese cabbage. Then a 50% of growth increase was achieved in 2 months (Hakalehto 2016) (See also Fig. 4.1.). The growing trend of antibiotic resistance strains in food is also another area of concern (Hakalehto, 2015c; Adusei-Mensah et al., 2021).

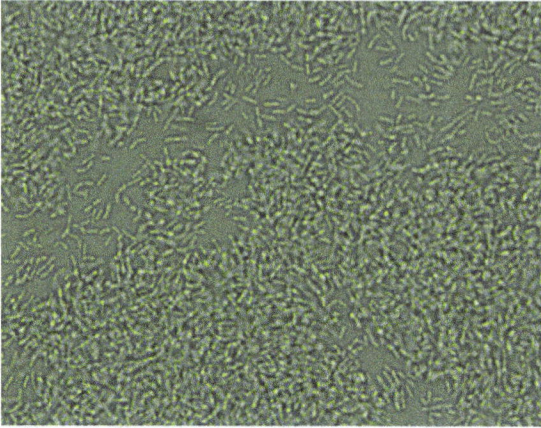

Figure 4.9: Phase contrast microscopy of a week-old LAB colony (1,000× enlarged). After a few days, millions of LAB cells were ready to initiate the fermentation process as an inoculum. Photo: Frank Adusei-Mensah, Finnoflag laboratory, Siilinjärvi Finland, 2021.

For example, lactic acid from LAB has been used to prolong shelf life of milk, cheese, yogurt, wine, beer, bread, sauerkraut, soy sauce, vegetables, fish, and fermented meats (Ross et al., 2002). Acetic acid from *Acetobacter aceti, A. xylinum*, and *A. ascendens* has been used similarly for vinegar production by suppressing the growth of undesired organisms with the added distinctive taste and flavour. Several preservatives can be produced by micro-organisms (Immonen et al., 2016). These compounds could also be beneficial for health, as like quercetin and other flavonoids.

4.11 ABOWE biorefinery project as an example of circular economy opportunities for recycling food waste and side streams

To gain the full benefit of the reuse of food residues or to get material and product loss in storage compensated, microbiological biorefinery approach offers a safe and proven solution (Hakalehto and Jääskeläinen, 2017).

For example, an average European-wide yearly recycled biowaste collection rate of 150 kg per citizen has been anticipated to form an additional stream of 30 million tons of surplus raw material annually (den Boer et al., 2020). One important option for the utilization of this vast asset would be its partial usage as complementing food or feed production. This could take place hygienically by microbial and enzymatic processes as tested, for example, in Poland, Finland, and Sweden during the ABOWE pilot study (den Boer et al., 2016a, b; Hakalehto et al., 2016a, b, c, Schwede et al., 2017)

An experimental biorefinery project was carried out by six European Union Baltic Sea region countries: Germany, Poland, Lithuania, Estonia, Sweden, and Finland. In ABOWE December 2012—December 2014, two microbial technologies were piloted and tested in semi-industrial mobile pilot plants (Figure 4.10). One technology was Pilot A Novel Biorefinery, designed, engineered, and constructed within ABOWE project. This biorefinery concept of bioengineering with mixed microbial strains for the simultaneous production of energy and chemicals is developed by Adj. Prof. Elias Hakalehto, Finland. This semi-industrial biorefining unit was constructed by the Savonia University of Applied Sciences together with tens of subcontractors and the key technology provider, Finnoflag Oy. Another technology was Pilot B based on a German dry fermentation process.

Figure 4.10: ABOWE Pilot A: Novel Biorefinery. The mobile unit is a pocket-sized factory in a 12 m sea freight container (on the right). Photograph was taken during the presentation of the biorefinery pilot at the Finnish Science Centre Heureka, Vantaa, April 2014, prior to the transportation of this satellite connected experimental station to Poland for the tests with potato waste. See the text. Photo: Ari Jääskeläinen.

ABOWE testing periods with Pilot A gave a proof of concept with different biomass waste materials in Finland, Poland, and Sweden that the biorefinery pilot plant was capable of functioning as an upstream biorefinery for all kinds of organic wastes. The products are biofuels (such as ethanol, butanol, hydrogen), bioenergy, organic platform chemicals, and organic fertilizers. These can be fabricated with the help of micro-organisms in an economically viable way (Hakalehto 2015e, Hakalehto et al.,

2016b). In Finland, dewatered wastewater treatment sludge from a carton board factory was used as substrate (Hakalehto et al., 2016a). In Poland, the products were short- and medium-chain volatile fatty acids (VFAs) and as a substrate there was used kitchen biowaste from the restaurants, as well as potato peels coming from a potato chip factory (den Boer et al., 2016a,b and 2020). In Figure 4.11, the presented formation of gaseous products in ABOWE Polish test run 10 indicates the high yield of biohydrogen. The yields and productivities of the Polish tests correspond to industrially feasible production levels of the liquids, such as butyric and propionic acids, valeric acid, and 2,3-butanediol. Interestingly, biohydrogen production in association with this process is a lucrative additional option. The highest glucose conversion level reached 0.81 mol/mol of glucose. Wroclaw University of Technology group, under the leadership of Assistant Professor Emilia den Boer, and involving several experts as well as over 20 students, designed the experiment with Dr. Elias Hakalehto and used the ABOWE pilot plant together with Finnoflag Oy staff. It was an interesting experiment to see how professionals and students from different countries could cooperate in using the novel bioreaction system packed with various technological solutions.

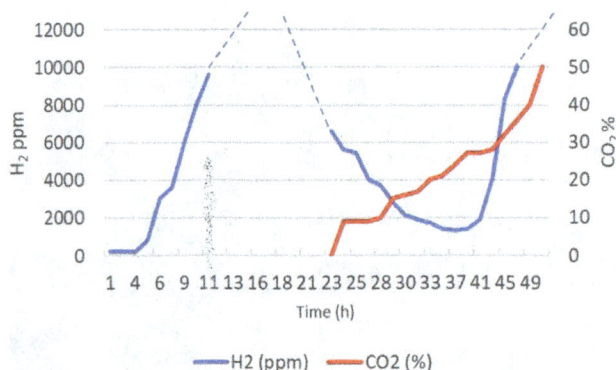

Figure 4.11: Production of Hydrogen (H_2) and carbon dioxide (CO_2) during the ABOWE experiment in Poland. The maximal measurable level of H_2 detection with the available sensor was 10,000 ppm. The first phase of Hydrogen production was initiated after short lag phase in about 5 h. There was a break in the process control between the time points of 11 and 23 h. It is noteworthy that these levels of gas production were measured from the insert carrier gas flow. The current result illustrates the potential for collecting "biohydrogen" as a side product from the biorefinery processes (modified from den Boer et al. (2016b)). Similar, relatively high production rate for Hydrogen, has been detected as a side activity in Finnoflag Oy's industrial projects as well as in the ABOWE pilot runs also in Finland and Sweden and the pilot tests in Hiedanranta, Tampere (see the text). Since this type of hydrogen collection is not requiring much energy input, we call the resulting energy gas "Biohydrogen" instead of somewhat misleading "Green Hydrogen" concept.

In Sweden, the tests were carried out by Mälardalen University, Finnoflag Oy, and Savonia University of Applied Sciences, from Kuopio, Finland. As the raw materials or substrates, the combined wastes from ecological chicken farm Hagby's Gardsfagel AB and their slaughterhouse were used for the process. During this experimentation, it turned out that it was possible to produce gases and chemical goods from the carboxylates as well as from the more challenging protein and lipid-containing wastes. Moreover, these promising results were obtained from a substrate mix of abattoir wastes with manure and wood chips (Hakalehto et al., 2016b, Schwede et al., 2017).

Biogas production test runs with Pilot B, in turn, were providing a proof of concept in Lithuania (using manure from a small farm), Estonia (using manure from a large farm), and Sweden (using sorted municipal solid waste at a waste management center). Professor Thorsten Ahrens from Ostfalia University of Applied Sciences, Germany, led these experiments in collaboration with the project partners in each EU Baltic Sea country (Freidank et al., 2014). It is highly important that in the context of sustainable food production, to process all the residual fractions into side products.

These kinds of holistic technology developments promote circular economy activities in agriculture, waste treatment sites, and industry (Figure 4.12). In the biorefinery field, the various processes and the corresponding material flows could form a network. This reflects the nature's way of operation, as there are no landfills, but all the materials are recycled, and energy is incessantly flowing. Residues from any process could be linked with other streams for new solutions. Furthermore, the leftovers from this network could be used in agriculture as organic fertilizers and hence directed back to circulation. Also, the gaseous flows could be redirected to the process, most importantly with the assimilation and reuse of organic carbon. This could mitigate the climate effects by replacing the incineration, or by combining the waste combustion with waste recycling and other processes (Hakalehto and Jääskeläinen, 2017). Such an overall concept could be effectively used for an educational strategy in implementing sustainable biorefineries in higher education (Jääskeläinen and Hakalehto, 2018).

In each region, business models were tentatively formed, and introductory seminars were held for possible stakeholders, based on the results from testing periods. Moreover, for example, the feedstock potentials and climatic impacts of the new processes from the point of view of each region and country were analyzed with a regional model developed at the University of Eastern Finland (Huopana et al., 2014). The process control and adjustment were described by Jääskeläinen et al. (2016).

ABOWE testing periods expressed the potential of holistic planning in biotechnological waste utilization and bioprocess design. For instance, the Swedish test run results confirmed that the Pilot A type of biorefinery's process broth could be efficiently used as raw material in the Pilot B type of dry digestion biogas production. When chicken manure that had been biorefined in Pilot A was further used as raw material for Pilot B, biogas production increased tenfold (Schwede et al., 2017).

Figure 4.12: Elias Hakalehto, Adj. Prof., PhD, inside the ABOWE movable pilot plant in 2014. This biorefinery unit contained four tanks: 1. homogenizer, 2. hydrolyzer, 3. bioreactor (patented), and 4. product collection tank, as well as the computer (or satellite) control unit and a small laboratory. Photo: Ari Jääskeläinen. "Future biorefineries are industrial fields, where side streams from industries and municipalities are treated into useful products in successive process solutions. All waste is then integrated as a raw material into them" (statement by Elias Hakalehto, 2014).

This occurred in a remarkably shorter time than the chicken manure dry digestion in Pilot B took place without Pilot A biorefining first (Dahlquist, 2016, Hakalehto et al., 2016c). Taking into account the types of fractions, solid fractions could be utilized as organic fertilizer, or composted or combusted (if no other means available). At least part of the gaseous emissions could be redirected into the biorefinery, enabling the utilization of their energies. As an example, the carbon oxides and some volatile nitrogen compounds could hence be bound into the products in the Pilot A type of biorefinery (Hakalehto et al., 2016b).

In the meat industry, it is always of high priority to monitor and confine the zoonotic and other hazardous micro-organisms, thus preventing their risky distribution or emission. Organized exploitation of the side streams or slaughterhouse wastes helps in the hygienization of the biomasses, processes, and the entire field of industry (Hakalehto et al., 2016b, c). Later on, after the conclusions and teachings of the ABOWE, the combination of sucrose industry with abattoir wastes was used for the production of mannitol (Hakalehto et al., 2016d). The usability of mannitol is a promising and versatile issue, as this sugar-alcohol could be used as low-caloric sweetener in the food industry, in the production of sweets and caramels, as well as an excipient for the pharmaceuticals (Ohrem et al., 2014).

Based on samples from the test runs from three countries, Ostfalia University of Applied Sciences designed downstream processing for non-gaseous products (Freidank and Ahrens, 2014), especially for 2,3-butanediol, which as a platform chemical was one of the target products in the Pilot A tests. This four-carbon substance, 2,3-butanediol is a non-toxic chemical commodity, which is widely used as an additive in many products. It was effectively produced from potato waste by microbes (Hakalehto et. al., 2013).

Potential other raw materials for the Pilot A type of biorefinery are the waste biomasses from agriculture and forestry. The major novelties of the Finnoflag biorefinery process are the enhanced productivity and wide product repertoire, both achieved by special innovations. As higher product levels can be attained faster (within, e.g., a couple of days only, instead of many weeks as in standard biogas processes), the plant size can be smaller, enabling lower investment. This faster throughput reactor and process solution could be used both in batch or fed-batch types, as well as in continuous bioprocesses. Moreover, the downstream processing of the products (separation, purification, etc.) can be done with lower costs, as end product concentrations can be increased, and the process throughput time shortened.

4.11.1 Some biochemicals as products for food-related uses from the microbiologically processed side streams

Several micro-organisms are very beneficial for the industries, also for food-derived, food-related products. Some examples of potential chemical products are presented here. Their manufacturing has been tested in the above-mentioned pilot runs. It is noteworthy, that the experiments were semi-industrial pilot trials, but they also could be extended to industrial production scale as numerous parameters and process principles were provisionally tested. Some ideas of applications are listed below. Several important short-chain fatty acids (SCFAs) are considered here as examples, since they form the potential for recycled compounds, for example, for food industries (den Boer et al., 2020). They are often produced in anoxic conditions. Acetic acid can be manufactured both in aerobic and anaerobic processes. Valeric acid could be condensed from acetic and propionic acids. Fermentation is a viable process alternative in all cases. For citric acid production, anaerobic process by bacteria or molds is often more feasible than chemical synthesis.

4.11.1.1 Acetic acid and propionic acid

Acetic and propionic acids are carboxylic acids that are found naturally in various organic sources. In winemaking, vinegar is formed, in which product acetic acid is the primary ingredient. In industry, acetic acid is a significant organic chemical. The estimated global demand for acetic acid is over 6.5 million tons, and it is projected to grow 3–4% annually (Xu et al., 2019). The global acetic acid market size was valued at USD 8.92 billion in 2019 (Grand View Research, 2020). Currently, the dominant technology for the industrial production of acetic acid is a chemical synthetic method from oil. Methanol carbonylation is the most common process, accounting for over 65% of the global production capacity (Xu et al., 2019). However, microbial fermentation is also a remarkable source. It represents a more sustainable process thinking, which also offers increasingly climate-friendly technological solutions.

The industrial production of propionic acid takes place almost completely from petrochemical sources. From the 20th century onward, numerous pilot production plants have been engineered and fabricated for propionic acid production via fermentation. However, these endeavors have not led to industrialization, mainly due to the too costly acid separation from the fermentation broth. The acid's concentration in the fermentation broth is rather low, and between the acid and the water there is only a marginal volatility gap. The biorefinery strategy, that is, involving renewable substrate sources for producing bulk products, has been exercised in the biofuels supply, for example, in the production of biodiesel or fuel ethanol. Moreover, bio-based chemicals have been produced (e.g., citric acid and glutamate). The same trend is also anticipated for producing acetic and propionic acids (Deppenmeier et al., 2002). The use of biological processes makes the chemical industries increasingly sustainable.

Acetic acid is an important chemical reagent and industrial chemical, which is used for several purposes (Deppenmeier et al., 2002):
– the production of polyethylene terephthalate for soft drink bottles,
– cellulose acetate for photographic film,
– polyvinyl acetate for wood glue,
– synthetic fibers and fabrics,
– households as diluted acetic acid in descaling agents,
– food industry as a food additive (code E260).

Prominent players in the acetate market include with the headquarter country mentioned (Grand View Research 2020):
– Eastman Chemical Company (USA),
– British Petroleum (UK),
– LyondellBasell (the Netherlands),
– Celanese Corporation (USA),
– Gujarat Narmada Valley Fertilizers & Chemicals (India),
– Helm AG (Germany),
– Pentoky Organy (India),
– Dow Chemical (USA),
– Indian Oil Corporation (India).

The preservative use of ammonium, potassium, sodium, and calcium salts of propionate is an important application in the production of food and animal feed. Propionic acid as a chemical intermediate is a significant factor in the synthesis of cellulose fibers, pharmaceuticals, perfumes, and herbicides (Xu et al. 2019).

Currently, propionic acid uses are segmented into various fields, according to (Xu et al., 2019):
- about 45% animal feed and grain preservatives,
- 21% food preservatives,
- 19% herbicides,
- 11% cellulose acetate propionate (CAP), and
- 4% miscellaneous uses.

Moreover, propionic acid and its derivatives are also used for several other purposes (Xu et al., 2019):
- pharmaceuticals production,
- artificial fruit flavours (e.g., geranyl propionate and citronellyl),
- plasticizers (e.g., phenyl propionate and glycerol tripropionate).

Global propionic acid market demand was almost 400,000 tons in 2013 and was then expected to exceed 470,000 tons in 2020 (Grand View Research, 2015). The global propionic acid market is expected to grow at a Compound Annual Growth Rate (CAGR) of over 5% during 2019–26 (Mordor Intelligence, 2020). It is projected to reach the revenue of USD 2 billion by 2024 (Market Research Future, 2021). The propionic acid market is highly consolidated in nature. Some of the largest producers of propionic acid are (Mordor Intelligence, 2020):
- BASF SE (Germany),
- Eastman Chemical Company (USA),
- Dow Chemical (USA),
- Perstorp (Sweden),
- DAICEL CORPORATION (Japan).

Previously acetic acid was often obtained from natural carbohydrates via ethyl alcohol's biochemical oxidation and wood's destructive distillation. Vinegar can also be produced from a large spectrum of raw materials using micro-organisms. Some examples are: grapes, apples, peaches, oranges, pineapples, pears, and other fruits. Moreover, vinegar could be obtained from sugar-based feedstock molasses as well as from grains such as malt and whey. "Vinegar" refers to the product of the acetic fermentation of ethanol from one of the substrates listed above. Nowadays, the biological route covers only approximately one-tenth of the global production but is still significant for producing vinegar. The reason is that laws require that vinegar used in foods must be of biological (natural) origin (Xu et al., 2019). However, by process improvement and by such novel technologies as the Finnoflag biorefining, it is possible to reach unprecedented productivities, which could make the biological route to acetate, and to many other products of microbial metabolism more economically lucrative, and eventually the first choice for the large-scale sustainable industries.

Acetic acid can be produced by both aerobic and anaerobic bioproduction assays (AAB) (Xu et al., 2019). AAB are used to oxidize ethanol to produce acetic acid in aerobic conditions. AAB are currently classified into 10 genera and 44 species:
- *Acetobacter* (16 species),
- *Gluconobacter* (5 species),
- *Acidomonas* (1 species),
- *Gluconacetobacter* (15 species),
- *Asaia* (3 species),
- *Kozakia* (1 species),
- *Saccharibacter* (1 species),
- *Swaminathania* (1 species),
- *Neosaia* (1 species), and
- *Granulibacter* (1 species), in the family Acetobacteraceae as a branch of the acidophilic bacteria in the alpha-subdivision of the Proteobacteria (Xu et al., 2019).

If the UMC approach by Finnoflag Oy could be further developed, the use of microbial communities could consequently increase for acetic acid production.

In acetic acid's anaerobic production process, in turn, *Clostridium thermoaceticum* (in 1994, renamed as *Moorella thermoacetica*) has been used to elucidate the acetic acid's homofermentation. This reaction converts 1 mol of glucose into 3 mol of acetic acid. Among all the homoacetogens known so far, only *M. thermoacetica* has been previously considered to have potential for industrial application. Therefore, it is the single one that has been comprehensively researched for the anaerobic fermentation of acetic acid so far. As the substrate, glucose, fructose, CO, xylose, lactate, and milk permeate have been used (Xu et al., 2019).

Conventional aerobic vinegar fermentation has the drawback of low product yield and too high energy consumption. On the contrary, in anaerobic fermentation, the substrate carbon can be almost totally recovered in acetic acid. Hence, it might be regarded as the optimal microbial process for the industrial production of acetic acid. However, the commercialization of the process has not realized till now. The two main issues to be solved are (Xu et al., 2019):
- when the concentration of acetate is high, this inhibits acetogens, and
- acidic conditions are not suitable for the growth of acetogens.

Another hindrance in the industrialization of acetogenesis is the fact that sugar polymers (e.g., cellulose) can be degraded by just a few acetogens. Co-cultures of cellulolytic *Clostridium thermocellum* and the thermophilic acetogen *Thermoanaerobacter kivui* are capable of producing acetate from cellulose, with almost total cellulose-derived carbon recovery (Svetlitchnyi, et al., 2013). Also, corresponding results have been received with co-culturing cellulolytic strain of *Ruminococcus albus* and the unclassified coccobacillus (Miller and Wolin, 1995).

Propionibacterium species are the ones that have been the most comprehensively studied ones for producing propionic acid. Hence, the United States Food and Drug Administration (FDA) has granted them the Generally Recognized as Safe (GRAS) classification (Xu et al., 2019). Also, several other anaerobic bacterial genera, for example, *Selenomonas, Veillonella*, and *Clostridium*, and especially the species *Clostridium propionicum*, form propionic acid as the primary fermentation product. Many strains utilize multiple sugars as substrates. *Veillonella parvula* can use pyruvate, lactate, and succinate, for example (Balamurugan et al., 1999).

Acetic acid and propionic acid separation from water has been an extensive target for research, and the following technologies have been developed for this purpose (Xu et al., 2019):

- solvent extraction,
- fractional distillation,
- azeotropic dehydration distillation,
- combination of the above methods,
- extractive distillation, and
- carbon adsorption.

Nowadays, major bulk producers of acetic or propionic acid are not exploiting fermentations for production, so the markets are primarily supplied petrochemically. Depending on the product standards, either an aerobic or anaerobic process is used for the fermentative production of acetic acid. The aerobic process, for example, is the primary one for vinegar production. In acetic acid production, in case the acid concentration in the broth being more than 5%, the energy need is lower in anaerobic acidogenesis (Xu et al., 2019). This opens room for approaches like the Finnoflag method for using microbial communities in the biorefineries.

Acetic acid acts as an inhibitor for the rate of acid production in both processes. Therefore, simultaneously removing acid from the process would enhance the substrate conversion to acids. This can be done by including a subsystem outside the fermentor, such as a membrane separator or a solvent extractor (Xu et al., 2019).

Anaerobic fermentation with *Propionibacterium* sp. and related strains mainly form propionic and acetic acids. The propionic acid production, however, is low because byproduct and propionic acid formation strongly inhibit cell growth and the fermentation itself. Propionic acid is highly needed for the natural preservation of foods and grains. This has triggered the development of novel fermentation processes. The goal is to accomplish the improved production of propionic acid from cheap biomass and food processing wastes. New viewpoints to process engineering, metabolic engineering, and genetic engineering levels have been studied for enhanced propionic acid production by *P. acidipropionici*. Revealing the mechanism for controlling the fermentation of propionic acid, and the mutants formed during research presumably enables the development of a viable bioprocess for producing propionic acid from various renewable substrates (Xu et al., 2019). In the slaughterhouse experiment by Finnoflag Oy, it was evidenced that in mixed cultures on abattoir wastes, the lactic acid is eventually converted into other SCFA's, such as propionic, acetic, and butyric acids (Hakalehto, 2015d).

4.11.1.2 Valeric acid

Valeric acid (valerate, pentanoic acid) is a low-molecular-weight colorless and oily carboxylic acid compound with an unpleasant odour. However, it can be refined into fine fragrances. It is primarily used for (Allied Market Research, 2021; Expert Market Research, 2021a):
-   the synthesis of its esters,
-   agricultural chemicals such as pesticides,
-   perfumes,
-   cosmetics products,
-   active pharmaceuticals ingredient intermediates,
-   laboratory, research, and development work.

Valeric acid has several additional applications, including
-   plasticizers,
-   ester-type lubricants, and
-   vinyl stabilizers.

Some esters of valeric acid such as pentyl valerate and ethyl valerate have been used as food additives, such as flavours (Allied Market Research, 2021).

Valeric acid could also be used as a platform chemical from waste utilization bioreactions. For example, 2,3-butanediol could also be used as a substrate for 1,3-butadiene (leading to synthetic rubber and plastic monomers). Valeric acid value could be as high as up to 5,000–7,000 € per metric ton (Hakalehto et al., 2016b).

The global valeric acid market is projected to grow at a CAGR of 5.3% during 2021–26 (Expert Market Research, 2021a).

The key producers in the global valeric acid market are, for example (Allied Market Research, 2021; Expert Market Research, 2021a):
-   Dow Chemical (USA),
-   Perstorp (Sweden),
-   Sisco Research Laboratories (India),
-   Sigma Aldrich (USA),
-   Central Drug House (India),
-   Otto Chemie Pvt. Ltd. (India),
-   Yufeng International Co., Ltd. (China).

Valeric acid is now being manufactured from bio-based sources (Allied Market Research, 2021). Valeric acid is microbiologically formed as a condensate of acetate and propionate (Hakalehto et al., 2016b). However, the major restraint for the global valeric acid market is its unpleasant odour and availability of other alternatives present in the market such as butyric acid (Allied Market Research, 2021). However, valeric acid was produced from mixed wastes during a pilot experiment, and it could also be converted into more pleasantly smelling fragrance compounds (Schwede et al., 2017).

4.11.1.3 Butyric acid

Butyric acid, also known as butanoic acid, refers to a fatty acid that occurs in the form of esters in plant oils and animal fats (Global Market Insights, 2021). It is also a common product of many anaerobic bacterial fermentations. Butyric acid is a carboxylic acid, and its salts and esters are called as butyrate. It has an unpleasant smell and acrid taste, with a sweetish aftertaste. The acid is an oily colourless liquid that is soluble in water, ethanol, and ether and can be separated from an aqueous phase by saturation with salts such as calcium chloride (Xu and Jiang, 2019).

Butyric acid market size exceeded USD 175 million in 2020 and is estimated to grow at over 13.2% CAGR between 2021 and 2027. By 2027, it is projected to grow up to USD 405 million (Global Market Insights, 2021).

Butyric acid application areas are (Global Market Insights, 2021):
- pharmaceuticals,
- biofuel,
- food additive and flavouring,
- animal feed,
- cosmetic,
- plasticizer,
- leather tanning,
- and so on.

Butyric acid has many important applications in food production, pharmaceutical industries, cosmetics, and chemical industries (Playne, 1985).

In food industry, butyric acid has an increasing usage as a food additive and for flavouring purposes. Renewable butyric acid is a preferable choice in food and beverage industry for these uses (Global Market Insights, 2021). Butyric acid is used for providing butter-like taste for food flavours, and its esters are popular as additives to increase fruit fragrance.

Butyric acid is one of the main energy sources for human body, as it is one of the SCFAs generated by bacterial fermentation of dietary fibers in the colon. Butyric acid is also marked as a suppressor of colon cancer (Playne, 1985). Its production by butyric acid clostridia is enhanced by carbon dioxide emission of intestinal lactic acid bacteria (Hakalehto and Hänninen, 2012).

Butyric acid's biological effects have been intensively researched, including therapeutic effect for hemoglobinopathies, cancer, and gastrointestinal diseases. Butyric acid derivatives have also been developed for producing antithyroid and vasoconstrictor drugs, and it can be used in anesthetics (Playne, 1985).

Butyric acid has vast demand in the pharmaceutical industry as an intermediate. In addition, health issues like obesity, diabetes, anxiety, and other chronic symptoms or diseases, as well as emerging new healthcare regulations, are both promoting its market growth (Global Market Insights, 2021).

In North America, the consumption of butyric acid in the pharmaceutical industry is expanding due to its use for reducing body weight. Butyric acid controls the balance between breaking down the fats on one hand, and the fatty acid synthesis on the other. Consumers are paying effort on maintaining fitness and controlling their weight, as an outcome from changes in lifestyles and ways of eating. This has led to various health problems, for example, gastric and digestive, and irritable bowel syndrome (Hakalehto, 2020). Butyric acid helps to keep the gut lining healthy and sealed, hence enhancing the health of the gut (Global Market Insights, 2021).

The cosmetic application of butyric acid is increasing as methyl butyrate is used in the perfume additives. The rise of people's living standards has expanded the demand for luxurious goods such as cosmetics. Moreover, major producers are putting effort in the research and development activities to bring to market, for example, soaps, shower gels and others. Consumer preferences have changed, and millennials are willing to try new, more exotic, and innovative products such as perfumes or other cosmetic products. They are willing to pay premium prices more readily (Global Market Insights, 2021).

Within the chemical industry, butyric acid is used for producing thermoplastics cellulose acetate. In addition, glycerol tributyrate and other esters also have a significant share in the plastic materials (Playne, 1985).

Butyric acid is a precursor for biofuels. Alternative fuel sources have been in the focus for research and development, because of the increasing price of petroleum and expanding demand for the sources of clean energy. Biofuels bring along environmental benefits, such as reduction of greenhouse emission and an option to replace gasoline. Biobutanol provides clear benefits in comparison with ethanol regarding the fuels for traffic (Xue et al., 2017). Moreover, bio-butanol is used as industrial solvents, too (Global Market Insights, 2021).

"Another encouraging example was the invention in the early days of biotechnological microbiology. During the years 1911–15, Sir Winston Churchill was the First Lord of Admiralty in Great Britain. He invited into his office a young scientist from East Europe, Dr. Chaim Weizmann, who later became the first president of Israel. Weizmann described in his autobiography his work during those times which had produced the finding of Weizmann's bacterium or *Clostridium acetobutylicum* and the so-called ABE fermentation (acetone, butanol, ethanol). Facilities for this work lacked at the university, where biochemistry did not form part of the curriculum at that time, while the study of bacteriology was confined to medical school. I began to pay frequent visits to Pasteur Institute in Paris, where I worked in the bacteriological and microbiological departments" (Weizmann, 1949). The anaerobic bacterial strains of several *Clostridium* sp. strains have been used in the recent biorefinery projects of Finnoflag Oy, such as in the European Union Baltic Sea biorefinery project ABOWE in 2012–14, which consisted of researchers from Poland, Germany, Sweden, Finland, Estonia, and Lithuania. The concept of non-aseptic cultures was introduced to produce valuable chemicals from the side streams. Later, this concept of "Finnoflag biorefining" has been piloted in several industrial settings including the "Zero waste from zero fibre" project in Tampere, Finland, where the lake bottom sediments of pulp and paper industries were exploited as raw materials for microbially produced biochemicals by the pilot studies conducted in 2018–19. The project consortium could turn all the side streams into useful products, including the mentioned chemicals, energy gases such as methane and hydrogen and organic fertilizers. These tests predicted the bright future with times when wastes will be most valuable and wanted raw materials. The bacterium, *Clostridium autobutylicum*, was effectively boosted by carbon dioxide in Finnoflag Oy's testing (Hakalehto, 2015e).

A century earlier, Sir Winston Churchill made a question to Chaim Weizmann: "Well, Dr Weizmann, we need thirty thousand tons of acetone. Can you make it?" Weizmann replied: "I do my work in the laboratory. I am not a technician; I am only a research chemist. But, if I were somehow able to produce a ton of acetone, I would multiply that by any factor you chose. Once the bacteriology of the process is established, it is only a question of brewing. I must get hold of a brewing engineer of one of the big distilleries, and we will set out the preliminary task. I shall naturally need my government's support to obtain the people, the equipment, the emplacements, and the rest of it. I myself cannot even determine what will be required." – This conversation led to an establishment of an entire field of industries which also helped in setting up the production of antibiotics some decades later when the time was right (Hakalehto, 2021c).

Concerning animal feed application, animals' infectious diseases are increasing. Butyric acid–based feed additive aids in protecting broiler chicken from *Salmonella enterica* serovar Enteritidis infection (Van Immerseel et al., 2005). Moreover, butyric acid helps in improving intestinal health and performance in swine and poultry. Increasing per capita income in India and China as well as improving lifestyle have led to a higher demand for top quality meat, resulting in growth in the livestock industry and thus further increase in demand for butyric acid (Global Market Insights, 2021).

Worldwide butyric acid business is fragmented, and major companies are, for example (Global Market Insights, 2021):

- OXEA GmbH (Germany),
- Beijing Huamaoyuan Fragrance Flavor Co., Ltd. (China),
- Perstorp Holding AB (Sweden),
- Dmitrievsky Chemical Plant (Russia),
- Eastman Chemical Company (USA),
- Tokyo Chemical Industry Co., Ltd. (Japan),
- Blue Marble Biomaterials (USA),
- Snowco Industrial Co., Ltd. (China),
- Alfa Aesar (USA).

Moreover, there are other manufacturers that are highly interested in the market (Global Market Insights, 2021).

Butyrate is nowadays produced industrially mainly via petrochemical routes through the oxidation of butyraldehyde that is obtained from propylene (Xu and Jiang, 2019).

Biotechnological production of butyric acid has a relatively low productivity and butyrate concentration in the fermentation broth. Hence, this bioprocess is not commercially viable, so far. However, as food and pharmaceutical manufacturers prefer biologically produced food additives or pharmaceutical products, improvements in the economics and efficiency of butyrate fermentation process are necessary. The substrate cost is commonly a substantial part of the total production costs. Hence, there is a strong need for low-grade biomass as the feedstock. This would lead to industrially economic production of butyric acid via the fermentation route (Xu and Jiang, 2019). Also, such approaches as the Finnoflag biorefinery technology could increase the productivity of this compound.

Renewable butyric acid is manufactured from natural sources such as sugar, corn husk, and other natural products (Global Market Insights, 2021). It is possible to extract butyric acid from butter, but this method is too costly. As carbon sources glucose, lactose from whey, sucrose from molasses, starch, potato wastes, corn meal, wheat flour, hydrolysate of corn fiber, cellulose, and xylose have been used. Butyric acid is the product of the butyrate metabolic pathway of the bacterial genera *Clostridium*, *Butyrivibrio*, *Butyribacterium*, *Sarcina*, and others.

The preferred strains belong to the genus *Clostridium*, which use two parallel metabolic pathways. Products of the first pathway, called acidogenesis, are acids, namely butyrate and acetate. Products of the second pathway, called solventogenesis, are solvents, namely butanol and acetone. A remarkable hindrance in butyrate's biotechnological production is the end-product inhibition. Butyric acid has a negative effect on the transmembrane pH gradient, whereas butanol affects membrane fluidity. This problem cannot be solved by existing fermentation techniques. Inhibition effects could be ceased or attenuated by online or in situ product removal (Xu and Jiang, 2019).

Extraction and pertraction are the most suitable methods for the butyric acid on-line recovery and *in situ* removal with collection from the process broth. Pertraction with a liquid membrane is more efficient than simple extraction. Fermentation processes combined with pertraction are performed in three liquid phases. The first is the fermentation broth, the second is the organic phase, and the third is the aqueous stripping solution. Liquid membrane (the organic phase) is simultaneously regenerated with aqueous stripping solution, in which the product can be concentrated. Adding a reactant or a carrier to the organic phase improves the distribution coefficient. These combined processes can be efficiently applied to a large group of biological products, including butyric acid (Xu and Jiang, 2019).

4.11.1.4 Citric acid

Citric acid, 2-hydroxypropane-1,2,3-tricarboxylic acid by its systematic name, is a natural and weak organic acid (Kirimura and Yoshioka, 2019; Expert Market Research, 2021b). Citric acid belongs to the most important bioproducts, thinking of production volume, availability, and usage opportunities (Kirimura and Yoshioka, 2019). It is commonly found in citrus fruits, such as lemons, limes, oranges, grapefruits, etc. (Expert Market Research, 2021b). Citric acid, an intermediate in aerobic metabolism through the tricarboxylic acid (TCA) cycle, is found everywhere in Nature. In TCA cycle, carbohydrates are oxidized to carbon dioxide for the cell metabolism (Kirimura and Yoshioka, 2019).

Citric acid and citrate, its salt form, are chemical products that have many uses in various industrial fields (Kirimura and Yoshioka, 2019). The main end-uses of citric acid are in food and beverages, household detergents and cleaners, as well as pharmaceuticals. In 2020, food and beverages accounted for most of the end-use (Expert Market Research, 2021b).

Citric acid is a popular food additive, and safe and non-toxic preservative, helping to stabilize and preserve food products. Citric acid is commonly used in beverages (Expert Market Research, 2021b):
- as an organic acidulant,
- for controlling the growth of organism,
- adjusting pH and enhancing flavours.

Besides acidulant use, citric acid has long been used in the setting of jams and in other ways in the confectionery industry due to its safety, pleasant acidic taste, and high water solubility (Kirimura and Yoshioka, 2019).

Citric acid is also used in bathroom cleaners and detergents due to its chelating effect with metals in hard water. It produces foam which helps in removing stain, without the softening need of water (Expert Market Research, 2021b) Moreover, citric acid has been used as a monomer for biodegradable polymers (Kirimura and Yoshioka, 2019).

Because of citric acid's safe and non-toxic nature, pleasant acidic taste, high water solubility, and chelating and buffering properties, it is also used in pharmaceuticals, including cosmetics (Expert Market Research, 2021b).

In the food and the pharmaceutical sectors, there is increasing demand for organic additives. Moreover, consumers prefer convenient and safe food products, such as citrate. The Joint FAO/WHO Expert Committee on Food Additives has granted citric acid as GRAS (Kirimura and Yoshioka, 2019).

These two driving forces are expanding the citric acid market. In addition, as the use of phosphates is restricted or prohibited by environmental regulatory authorities, its replacement by citric acid has also increased the demand (Expert Market Research, 2021b).

There is a steady increase in the annual worldwide production of citric acid: 0.9 million tons in 2000, 1.7 million tons in 2010, and 2.1 million tons in 2016 (Kirimura and Yoshioka, 2019). It reached 2.4 million tons in 2020 and is estimated to grow to 2.9 million tons by 2026. China's share of the total global production in 2020 was over 50%. The USA and Europe were the next largest production areas (Expert Market Research, 2021b).

The global citric acid market is fragmented, so there are several producers surrounding the market. The top producers are (Expert Market Research, 2021b):
- Weifang Ying Xuan (China),
- COFCO Biochemical (China),
- Lemon Bio-chemical (China),
- Jungbunzlauer (Switzerland),
- Tate and Lyle (UK),
- Cargill (USA).

Nowadays, citric acid is commercially fabricated via fermentation. Many microorganisms such as filamentous fungi, bacteria, and yeasts have been known to form citric acid and have been researched for this use. Among them, the filamentous fungi *Aspergillus niger* has been found to be the most efficient producer and is exclusively used as it has high citric acid productivity at low pH, and it does not secrete toxic byproducts, making it easy to handle (Kirimura and Yoshioka, 2019; Röhr, 1998; Röhr et al., 1996). As substrates, molasses and starch (hydrolysate or residue) are commonly used (Kirimura and Yoshioka, 2019).

Kirimura et al. (2011) have described that for citric acid manufacturing, there is an immediate need for developing a novel bioprocess, which should represent economically feasible and environmentally sustainable production technologies (Kirimura and Yoshioka, 2019).

Many studies regarding citric acid production by *A. niger* have brought about new facts. Moreover, there are still plenty of opportunities to improve the commercial production process specifically in the fermentation process and the product recovery (Kirimura and Yoshioka, 2019).

Generally, there are three methods for recovering citric acid from the broth or crude fermentation fluid:
– precipitation,
– solvent extraction, and
– separation by ion-exchange chromatography.

The two first ones are used industrially (Röhr et al., 1996; Röhr, 1998). These both require preprocessing, either by filtration or centrifugation, to remove mycelial debris and insoluble residues from the fluid. Precipitation is the classical method and mostly used, and applicable to all fermentation process types. It is done by adding calcium oxide hydrate (lime). The third method on the list above has also been experimented, and its commercial application is being tested (Kirimura and Yoshioka, 2019).

Moreover, citric acid is an intermediate in the TCA cycle, which is typically overproduced, and it remains still unclear why that is the case. Hence, there are many technical problems to be solved but also promises to be seen (Röhr et al., 1996; Röhr, 1998; Ruijter et al., 2002; Karaffa and Kubicek, 2003; Berovic and Legisa, 2004; Papagianni, 2007; Kirimura et al., 2011). Kirimura and Yoshioka (2019) consider bioimaging analysis and metabolic engineering by genome editing as very helpful techniques for fundamentally researching the metabolism of citric acid production. With the help of increasing basic understanding, the biochemical networks, as well as the regulatory mechanisms related to citric acid biotransfer and bioengineering solutions could be found. These could then offer a possibility for improving extant industrial strains and the production process of citric acid (Kirimura and Yoshioka, 2019). The different biochemicals and food components can be obtained from numerous plant materials and agricultural feedstocks (Hellstrand and Dahlquist, 2017). The versatility of biotechnical means is an important asset in the global change toward the sustainability in future. For example, it is also possible to use the same production units for different processes and goals, depending on the demand and need, as well as the availability of the raw materials.

4.12 Conclusions

Since we have learnt to think about microbes as dangerous pathogens, constituents of dirt, or as process contaminants, it is high time to see their potentials in food production, health maintenance, industrial usage, circular economy, as well as in the ecosystem engineering and other environmental uses. It has been well-known for many decades that natural microbes have a remarkable potential for producing chemical goods. Lately, metabolic engineering research has shown some promising results for process improvement. In the trials conducted and supervised by Finnoflag Oy, the applications of mixed communities, for example, the Undefined Mixed Cultures (UMC) have been successfully used for intensifying the process, and for increasing productivity and yield. Therefore, the organic side streams offer new materials which could be microbiologically converted in a sustainable, feasible, and socially acceptable way into such products as valuable chemicals, food substituents, energy gases, and organic fertilizers.

References

Adusei-Mensah, F., Hakalehto, E., Tikkanen-Kaukanen, C. (2021). Microbiological and Chemical Safety of African Herbal and Natural Products. Berlin, Germany: De Gruyter. 2021. Manuscript in print.

Allied Market Research (2021). Valeric acid market: Global opportunity analysis and industry forecast, 2020–2027. Available at: Valeric Acid Market Size, Share | Industry Growth & Forecast, 2027 (https://alliedmarketresearch.com) Referred to: 10. 4.2021.

Ansong, M. A. (2020). Contribution of Ghanaian traditional fermented foods towards sustainable development. Citinewsroom – Comprehensive News in Ghana. 2020. Available from: https://citinewsroom.com/2020/08/contribution-of-ghanaian-traditional-fermented-foods-towards-sustainable-development-article/

Balamurugan, K., Venkata Dasu, V., Panda, T. (1999). Propionic acid production by whole cells of Propionibacterium freudenreichii. Bioprocess Engineering, 20: 109–116.

Beckinghausen, A., Dahlquist, E., Schwede, S., Lindroos, N., Retkin, R., Laatikainen, R., Hakalehto, E. (2019). Downstream processing of biorefined lactate from lake bottom zero fiber deposit – A techno-economic study on energy-efficient production of green chemicals. 11th International Conference of Applied Energy, Västerås 12–16 August, 2019.

Beganovic, J., Pavunc, A. L., Gjuracic, K., Spoljarec, M., Suskovic, J., Kos, B. (2011). Improved sauerkraut production with probiotic strain lactobacillus plantarum L4 and Leuconostoc mesenteroides LMG 7954. Journal of Food Science, 76 (2): M124–M129. doi: https://doi.org/10.1111/j.1750-3841.2010.02030.x.

Berovic, M., Legisa, M. (2004). Citric acid production. Biotechnology Annual Review, 2004 (13): 303–343.

Cerimi, K., Akkava, K. C., Pohl, C., Schmitt, B., Neubanen, P. (2019). Fungi as source for new bio-based materials: A patent review. Fungal Biology and Biotechnology, 6: 17. doi: https://doi.org/10.1186/s40694-019-0080-y.

Ciriminna, R., Meneguzzo, F., Delisi, R., Pagliaro, M. (2017). Citric acid: Emerging applications of key biotechnology industrial product. Chemistry Central Journal, 11: 22.

Dahlquist, E. (2016). Environment friendly contribution of the microbes to energy conversion. In: Hakalehto, E. (ed.). Microbiological Industrial Hygiene. New York, NY, USA: Nova Science Publishers, Inc.

Den Boer, E., Den Boer, J. (2018). Environmental effects of the management of municipal waste, including the impact of organic recycling. In: Hakalehto, E. (ed.). Microbiological Environmental Hygiene. New York, NY, USA: Nova Science Publishers, Inc.

Den Boer, E., Lucaszewska, A., Kluczkiewicz, D., Lewandowska, D., King, K., Reijonen, T., Suhonen, A., Jääskeläinen, A., Heitto, A., Laatikainen, R., Hakalehto, E. (2016a). Volatile fatty acids as an added value from biowaste. Waste Management, 58: 62–69.

Den Boer, E., Łukaszewska, A., Kluczkiewicz, W., Lewandowska, D., King, K., Jääskeläinen, A., Heitto, A., Laatikainen, R., Hakalehto, E. (2016b). Biowaste conversion into carboxylate platform chemicals. In: Hakalehto, E. (Ed.). Microbiological Industrial Hygiene. New York, NY, USA: Nova Science Publishers, Inc.

Den Boer, E., Den Boer, J., Hakalehto, E. (2020). Volatile fatty acids production from separately collected municipal biowaste through mixed cultures fermentation. Journal of Water Process Engineering, 38: 101582.

Deppenmeier, U., Hoffmeister, M., Prust, C. (2002). Biochemistry and biotechnological applications of Gluconobacter strains. Applied Microbiology and Biotechnology, 60: 233–242.

Expert Market Research (2021a). Global valeric acid market outlook. Available at: https://www.ex pertmarketresearch.com/reports/valeric-acid-market Referred to: 10.4.2021.

Expert Market Research (2021b). Global Citric Acid Market to Reach 2.91 Million Tons by 2026. Press release. Available at: Global Citric Acid Market to Reach 2.91 Million Tons by 2026 (https://expertmarketresearch.com) Referred to: 1. 4.2021.

FAO (1998). Fermented Fruits and Vegetables. A Global Perspective. Edited by Battcock, M., Azam-Ali, S. FAO Agricultural Services Bulletin No. 134. Rome.

Freidank, T., Ahrens, T. (2014). Design of downstream processing of non-gaseous products from Finnoflag process. EU Baltic Sea Region Project ABOWE (Implementing Advanced Concepts for Biological Utilization of Waste). Report no: 3.9. Wolfenbüttel, Germany: Ostfalia University of Applied Sciences. Available at: www.abowe.eu

Freidank, T., Drescher-Hartung, S., Behnsen, A., Ahrens, T. (2014). Final output report – Comparison and conclusions of Pilot B operation in Lithuania, Estonia, and Sweden. EU Baltic Sea Region Project ABOWE (Implementing Advanced Concepts for Biological Utilization of Waste). Report no: 4.7. Wolfenbüttel, Germany: Ostfalia University of Applied Sciences. Available at: www. abowe.eu

Fu, C. Y., Huang, H., Wang, X. M., Liu, Y. G., Wang, Z. G., Cui, S. J., Gao, H. L., Li, Z., Li, J. P., Kong, X. G. (2006). Preparation and evaluation of anti-SARS coronavirus IgY from yolks of immunized SPF chickens. Journal of Virological Methods, 133: 112–115.

Global Market Insights (2021). Global butyric acid market. Available at: https://www.gminsights. com/industry-analysis/butyric-acid-market Referred to: 9.4.2021.

Goldberg, I., Rokem, J. S. (2009). Organic and fatty acid production, microbial. In: Schaechter, M. (ed.). Encyclopedia of Microbiology, Third ed. Oxford: Academic Press, 421–442. Available from https://www.sciencedirect.com/science/article/pii/ B9780123739445001565.

Gonzalez-Garcia, R. A., McCubbin, T., Navone, L., Stowers, C., Nielsen, L. K., Marcellin, E. (2017). Microbial propionic acid production Fermentation. Multidisciplinary Digital Publishing Institute, 3: 21.

Grand View Research (2015). Propionic acid market size worth $1.53 Billion By 2020. Available at: https://www.grandviewresearch.com/press-release/global-propionic-acid-market Referred to: 4.4.2021.

Grand View Research (2020). Acetic acid market analysis report. Available at: https://www.grand viewresearch.com/industry-analysis/acetic-acid-market Referred to: 4.4.2021.

Guzel-Seydim, Z. B., Kok-Tas, T., Greene, A. K., Seydim, A. C. (2011). Review: Functional Properties of Kefir. Critical Reviews in Food Science and Nutrition, 51 (3): 261–268.

Hakalehto, E. (ed.). (2012). Alimentary Microbiome – A PMEU Approach. New York, NY, USA: Nova Science Publishers, Inc.

Hakalehto, E. (2015a). Hygienic lessons from the dairy microbiology cases. In: Hakalehto, E. (ed.). Microbiological Food Hygiene. New York, NY, USA: Nova Science Publishers, Inc.

Hakalehto, E. (2015b). Hazards and prevention of food spoilage. In: Hakalehto, E. (ed.). Microbiological Food Hygiene. New York, NY, USA: Nova Science Publishers, Inc.

Hakalehto, E. (2015c). Antibiotic resistance in foods. In: Hakalehto, E. (ed.). Microbiological Food Hygiene. New York, NY, USA: Nova Science Publishers, Inc.

Hakalehto, E. (2015d). Microbial presence in foods and in their digestion. In: Hakalehto, E. (ed.). Microbiological Food Hygiene. New York, NY, USA: Nova Science Publishers, Inc.

Hakalehto, E. (2015e) Enhanced Microbial Process in the Sustainable Fuel Production In: Yan, J. (ed.). Handbook of Clean Energy Systems, United Kingdom: John Wiley & Sons

Hakalehto, E. (2016) The many microbiomes. In: Hakalehto, E. (ed.) Microbiological Industrial Hygiene. New York, NY, USA: Nova Science Publishers, Inc.

Hakalehto, E. (2018a). The effects of bioprocess scale, intracellular milieu and environmental parameters. In: Hakalehto, E. (ed.). Microbiological Environmental Hygiene. New York, NY, USA: Nova Science Publishers, Inc.

Hakalehto, E. (2018b). Modes and consequences of the microbial interactions. In: Hakalehto, E. (ed.). Microbiological Environmental Hygiene. New York, NY, USA: Nova Science Publishers, Inc.

Hakalehto, E. (2020). Current megatrends in food production related to microbes. Journal of Food Chemistry and Nanotechnology, 6 (1): 78–87.

Hakalehto, E. (2021a). Probiotics and prebiotics balance the food uptake and gut defences. 7th International Conference Food Chemistry & Technology (FCT-2021), scheduled for November 08-10, 2021; Paris, France.

Hakalehto, E. (2021b). Chicken IgY antibodies provide mucosal barrier against SARS-CoV-2 virus and other pathogens. IMAJ, 23: 208–211.

Hakalehto, E. (2021c). Past strive for safety innovations paving the way to sustainable health future. In: UVC-LED BLOG & NEWS. LED FUTURE. First published on 18th of February, 2021. Modified from the original blog article.

Hakalehto, E., Dahlquist, E. (2018). A microbiological approach to the ecosystem services. In: Hakalehto, E. (ed.). Microbiological Environmental Hygiene. New York, NY, USA: Nova Science Publishers, Inc.

Hakalehto, E., Hänninen, O. (2012) Gaseous CO2 signal initiates growth of butyric-acid-producing Clostridium butyricum in both pure culture and mixed cultures with Lactobacillus brevis. Canadian Journal of Microbiology, 58(7): 928–931.

Hakalehto, E., Jaakkola, K. (2013). Synergistic effect of probiotics and prebiotic flax product on intestinal bacterial balance. Clinical Nutrition, 32 (Supplement 1): S200–201.

Hakalehto, E., Jääskeläinen, A. (2017). Reuse and circulation of organic resources and mixed residues. In: Dahlquist, E., Hellstrand, S. (Eds.). Natural Resources available today and in the Future: How to Perform Change Management for Achieving a Sustainable World. Germany: Springer Verlag.

Hakalehto, E., Kuronen, I. (1997). Therapeutic and preventive method against harmful microbes. International Patent Application, 1997; WO 97/37636.

Hakalehto, E., Kuronen, I. (1998). A method for producing jelly sweets which contain antibodies. International Patent Application, 1998; WO 98/43610.

Hakalehto, E., Humppi, T., Paakkanen, H. (2008). Dualistic acidic and neutral glucose fermentation balance in small intestine: Simulation in vitro. Pathophysiology, 15: 211–220.

Hakalehto, E., Hell, M., Bernhofer, C., Heitto, A., Pesola, J., Humppi, T., Paakkanen, H. (2010). Growth and gaseous emissions of pure and mixed small intestinal bacterial cultures: Effects of bile and vancomycin. Pathophysiology, 17: 45–53.

Hakalehto, E., Vilpponen-Salmela, T., Kinnunen, K., Von Wright, A. (2011). Lactic acid bacteria enriched from human gastric biopsies. ISRN Gastroenterol 2011. Available from: https://www.ncbi.nlm.nih.gov/pmc/articles/PMC3168382/

Hakalehto, E., Heitto, A., Pesola, J. (2012). Fecal microbiological analysis in the health monitoring. In: Hakalehto, E. (ed.). Alimentary Microbiome – A PMEU Approach. New York, NY, USA: Nova Science Publishers, Inc.

Hakalehto, E., Jääskeläinen, A., Humppi, T., Heitto, L. (2013). Production of energy and chemicals from biomasses by micro-organisms. In: Dahlquist, E. (ed.). Biomass as Energy Source: Resources, Systems and Applications. London, UK: CRC Press, Taylor & Francis Group.

Hakalehto, E., Heitto, A., Niska, H., Suhonen, A., Laatikainen, R., Heitto, L., Antikainen, E., Jääskeläinen, A. (2016a). Forest industry hygiene control with reference to waste refinement. In: Hakalehto, E. (ed.). Microbiological Industrial Hygiene. New York, NY, USA: Nova Science Publishers, Inc.

Hakalehto, E., Heitto, A., Suhonen, A., Jääskeläinen, A. (2016b). ABOWE project concept and Proof of Technology. In: Hakalehto, E. (ed.). Microbiological Industrial Hygiene. New York, NY, USA: Nova Science Publishers, Inc.

Hakalehto, E., Heitto, A., Andersson, H., Lindmark, J., Jansson, J., Reijonen, T., Suhonen, A., Jääskeläinen, A., Laatikainen, R., Schwede, S., Klintenberg, P., Thorin, E. (2016c). Some remarks on processing of slaughterhouse wastes from ecological chicken abattoir and farm. In: Hakalehto, E. (ed.). Microbiological Industrial Hygiene. New York, NY, USA: Nova Science Publishers, Inc.

Hakalehto, E., Heitto, A., Kivelä, J., Laatikainen, R. (2016d). Meat industry hygiene, outlines of safety and material recycling by biotechnological means. In: Hakalehto, E. (ed.). Microbiological Industrial Hygiene. New York, NY, USA: Nova Science Publishers, Inc.

Hakalehto,E., Väätäinen, U., Heitto, A., Ikonen, M., Tuomainen, S., Pesola, J., Jaakkola, K., Hänninen, O. (2018). Microbial environment on our skin, epithelial surfaces and tissues links us with the surrounding milieu. In: Hakalehto, E. (ed.) Microbiological Environmental Hygiene. New York, NY, USA: Nova Science Publishers, Inc.

Halm, M., Lillie, A., Sørensen, A. K., Jakobsen, M. (1993). Microbiological and aromatic characteristics of fermented maize doughs for kenkey production in Ghana. International Journal of Food Microbiology, 19: 135–143.

Hatta, H., Tsuda, K., Ozeki, M., Kim, M., Yamamoto, T., Otake, S., Hirasawa, M., Katz, J., Childers, N. K., Michalek, S. M. (1997). Passive immunization against dental plaque formation in humans: Effect of a mouth rinse containing egg yolk antibodies (IgY) specific to Streptococcus mutans. Caries Research, 31: 268–274.

Heer, K., Sharma, S. (2017). Microbial pigments as a natural color: A review | international journal of pharmaceutical sciences and research. available from: https://ijpsr.com/bft-article/micro bial-pigments-as-a-natural-color-a-review/?view=fulltext

Hefetz, A., Blum, M. S. (1978). Biosynthesis of formic acid by the poison glands of formicine ants. Biochimica et Biophysica Acta (BBA). General Subjects, 543: 484–496.

Hell, M., Bernhofer, C., Pesola, J., Pesola, I., Hakalehto, E. (2015). Prevalence, detection and prevention of foodborne outbreaks related to large hospital kitchens. In: Hakalehto, E. (ed.). Microbiological Food Hygiene. New York, USA: Nova Publishers, Inc.

Hellstrand, S., Dahlquist, E. (2017). Natural resources available today and in the future – how to perform change management for achieving a sustainable world. In: Dahlquist, E., Hellstrand, S. (eds.). Biologic Resources. Cham, Switzerland: Springer International Publishing AG.

Hernández-Velasco, P., Morales-Atilano, I., Rodríguez-Delgado, M., Rodríguez-Delgado, J. M., Luna-Moreno, D., Ávalos-Alanís, F. G., et al. (2020). Photoelectric evaluation of dye-sensitized solar cells based on prodigiosin pigment derived from Serratia marcescens 11E. Dyes and Pigments, 177: 108278.

Ho, C. W., Lazim, A. M., Fazry, S., Zaki, U. K. H. H., Lim, S. J. (2017). Varieties, production, composition and health benefits of vinegars: A review. Food Chemistry: 1621–1630.

Hoarau, G., Mukherjee, P. K., Gower-Rousseau, C., Hager, C., Chandra, J., Retuerto, M. A., Neut, C., Vermeire, S., Clemente, J., Colombel, J. F., Fujioka, H., Poulain, D., Sendid, B., Ghannoum, M. A. (2016). Bacteriome and mycobiome infections underscore microbial dysbiosis in familial Crohn's disease. American Society of Microbiology, mBio, 7: 5.

Huopana, T., Niska, H., Kolehmainen, M., Jääskeläinen, A., Antikainen, E., Schwede, S., Thorin, E., Klintenberg, P., Hakalehto, E., Ahrens, T. (2014). Sustainability assessment of biorefinery and dry digestion systems. Case: Sweden. EU Baltic Sea Region Project ABOWE (Implementing Advanced Concepts for Biological Utilization of Waste). Report no: 2.15. Kuopio, Finland: University of Eastern Finland. Available at: www.abowe.eu

Immonen, A., Immonen, I., Hakalehto, E., Del Amo, E. (2016). Preservatives in ocular medications and nutraceuticals. In: Hakalehto, E. (ed.). Microbiological Industrial Hygiene. New York, NY, USA: Nova Science Publishers, Inc.

Immonen, M., Hakalehto, J.-P., Hakalehto, E. (2015). Trends towards clean and healthy nutrition. In: Hakalehto, E. (ed.). Microbiological Food Hygiene. New York, NY, USA: Nova Science Publishers, Inc.

Jääskeläinen, A., Hakalehto, E. (2018). Biorefinery education as a tool for teaching sustainable development. In: Leal, W. (Ed.). Implementing Sustainability in the Curriculum of Universities. Cham, Germany: Springer International Publishing.

Jääskeläinen, A., Rissanen, R., Jakorinne, A., Suhonen, A., Kuhmonen, T., Reijonen, T., Antikainen, E., Heitto, A., Hakalehto, E. (2016). How does modern process automation understand the principles of microbiology and Nature. 9th EUROSIM Congress on Modelling and Simulation (EUROSIM 2016) 12–16 September 2016 Oulu, Finland.

Kabak, B., Dobson, A. D. W. (2011). An introduction to the traditional fermented foods and beverages of Turkey. Critical Reviews in Food Science and Nutrition, (51): 248–260.

Kanelli, M., Mandic, M., Kalakona, M., Vasilakos, S., Kekos, D., Nikodinovic-Runic, J., Topakas, E. (2018). Microbial rroduction of violacein and process optimization for dyeing polyamide fabrics with acquired antimicrobial properties. Front Microbiol 9. Available from: https://www.ncbi.nlm.nih.gov/pmc/articles/PMC6048185/.

Karaffa, L., Kubicek, C. P. (2003). Aspergillus niger citric acid accumulation: Do we understand this well working black box?. Applied Microbiology and Biotechnology, 61: 189–196.

Khalifa, S. A. M., Elias, N., Farag, M. A., Chen, L., Saeed, A., Hegazy, M. F., Moustafa, M. S., Abd El-Wahed, A., Al-Mousawi, S. M., Musharraf, S. G., Chang, F. R., Iwasaki, A., Suenaga, K., Alajlani, M., Göransson, U., El-Seedi, H. R. (2019). Marine natural products: A source of novel anticancer drugs. Marine Drugs, 17 (9): 491. doi: 10.3390/md17090491. PMID: 31443597; PMCID: PMC6780632.

Kirimura, K., Honda, Y., Hattori, T. (2011). Citric acid. In: Moo-Young, M. (ed.). Comprehensive Biotechnology. London, UK: Elsevier.

Kirimura, K., Yoshioka, I. (2019). Citric acid. In: Moo-Young, M. (ed.). Comprehensive
Biotechnology. Elsevier Science & Technology.

Kivelä, J., Hakalehto, E. (2016). Fertilization uses of meat bone meal and effects on microbial
activity. In: Hakalehto, E. (ed.). Microbiological Industrial Hygiene. New York, NY, USA: Nova
Science Publishers, Inc.

Li, Q., Xing, J. (2015). Microbial succinic acid production using different bacteria species. In: Kamm,
B. (ed.). Microorganisms in Biorefineries Berlin, Heidelberg: Springer. Available from https://
doi.org/10.1007/978-3-662-45209-7_7.

Market Research Future (2021). Global Propionic acid market research report. Available at: https://
www.marketresearchfuture.com/reports/propionic-acid-market-1122 Referred to: 4. 4.2021.

Miller, T., Wolin, M. (1995). Bioconversion of Cellulose to Acetate with Pure Cultures of
Ruminococcus albus and a Hydrogen-Using Acetogen. Applied and environmental
microbiology, 61: 3832–5. 10.1128/AEM.61.11.3832-3835.1995.

Mooradian, A. D., Smith, M., Tokuda, M. (2017). The role of artificial and natural sweeteners in
reducing the consumption of table sugar: A narrative review. Clinical Nutrition ESPEN, 18: 1–8.

Mordor Intelligence (2020). Propionic acid market. Available at: https://www.mordorintelligence.
com/industry-reports/propionic-acid-market Referred to: 4.4.2021.

Nguyen, H. H., Tumpey, T. M., Park, H. J., Byun, Y. H., Tran, L. D., Nguyen, V. D., Kilgore, P. E.,
Czerkinsky, C., Katz, J. M., Seong, B. L., Song, J. M., Kim, Y. B., Do, H. T., Nguyen, T., Nguyen,
C. V. (2010, Apr 13). Prophylactic and therapeutic efficacy of avian antibodies against influenza
virus H5N1 and H1N1 in mice. PLoS One, 5 (4): e10152. doi: 10.1371/journal.pone.0010152.
PMID: 20405007; PMCID: PMC2854139.

Nilsson, E., Larsson, A., Olesen, H. V., Wejåker, P. E., Kollberg, H. (2008). Good effect of IgY against
Pseudomonas aeruginosa infections in cystic fibrosis patients. Pediatric Pulmonology, 43 (9):
892–899.

Ohrem, H. L., Schornick, E., Kalivoda, A., Ognibene, R. (2014). Why is mannitol becoming more and
more popular as a pharmaceutical excipient in solid dosage forms? Review Article.
Pharmaceutical Development and Technology, 19 (3).

Okabe, M., Lies, D., Kanamasa, S., Park, E. Y. (2009). Biotechnological production of itaconic acid
and its biosynthesis in Aspergillus terreus. Applied Microbiology and Biotechnology,
84: 597–606.

Papagianni, M. (2007). Advances in citric acid fermentation by Aspergillus niger: Biochemical
aspects, membrane transport and modeling. Biotechnology Advances, 25: 244–263.

Passow, U., Overton, E. B. (2021). The complexity of spills: The fate of the Deepwater Horizon oil.
Annual Review of Marine Science, 13: 109–136. doi: 10.1146/annurev-marine-032320-095153.
Epub 2020 Sep 21. PMID: 32956014.

Patra, J. K., Das, G., Paramithiotis, S., Shin, H.-S. (2016). Kimchi and other widely consumed
traditional fermented foods of Korea: A Review. Frontiers in Microbiology, 7. Available from
https://www.frontiersin.org/articles/10.3389/fmicb.2016.01493/full.

Proksch, G., Baganz, D. (2020). CITYFOOD: Research Design for an International, Transdisciplinary
Collaboration, Technology|Architecture + Design, 4(1): 35–43, DOI: 10.1080/24751448.2020.
1705714.

Playne, M. J. (1985). Propionic acid and butyric acids. In: Moo-Young, M. (ed.). Comprehensive
Biotechnology, Vol. 3. Oxford, UK: Pergamon.

Rhee, S. J., Lee, J.-E., Lee, C.-H. (2011). Importance of lactic acid bacteria in Asian fermented foods.
Microbial Cell Factories, 10: S5.

Röhr, M. (1998). A century of citric acid fermentation and research. Food Technology Biotechnology,
36: 163–171.

Röhr, M., Kubicek, C. P., Kominek, J. (1996). Citric acid. In: Rehm, H. J., Reed, G. (eds.). Products of Primary Metabolism. Biotechnology, Vol. 6, 2nd ed. Weinheim, Germany: Wiley-VCH.

Ross, R. P., Morgan, S., Hill, C. (2002). Preservation and fermentation: Past, present and future. International Journal of Food Microbiology, 79: 3–16.

Ruijter, G. J., Kubicek, C. P., Visser, J. (2002). Production of organic acids by fungi. In: Oseiwacz, H. D. (ed.). Industrial Applications. The Mycota, Vol. 10. Berlin, Germany: Springer-Verlag.

Saha, B. C., Nakamura, L. K. (2003). Production of mannitol and lactic acid by fermentation with Lactobacillus intermedius NRRL B-3693. Biotechnology and Bioengineering, 82: 864–871.

Samgina, T. Y., Vorontsov, E.A., Gorshkov, V.A., Hakalehto, E., Hanninen, O., Zubarev, R.A., Lebedev, A.T.(2012). Composition and antimicrobial activity of the skin peptidome of Russian brown frog Rana temporaria. J Proteome Res, 11(12):6213–22. doi: 10.1021/pr300890m. Epub 2012 Nov 13. PMID: 23121565.

Samgina, T. Y., Tolpina, M. I., Hakalehto, E., Artemenko, K. A., Bergquist, J., Lebedev, A. T. (2016). Proteolytic degradation and deactivation of amphibian skin peptides obtained by electrical stimulation of their dorsal glands. Analytical and Bioanalytical Chemistry, 408: 3761–3768.

Sansonetti, P. (2019). The gut microbiome, from health to disease: Revisiting the Koch's postulates. Public conference in Helsinki on 25 th June.

Saranraj, P. (2019). Lactic Acid Fermentation – an overview | ScienceDirect Topics. Innovations in Traditional Foods. Available from: https://www.sciencedirect.com/topics/agricultural-and-biological-sciences/lactic-acid-fermentation

Sauramäki, J., Hakalehto, E. (2015). Catering services and hygienic food deliveries (Posti Ltd, Helsinki, Finland, and others). In: Hakalehto, E. (ed.). Microbiological Food Hygiene. New York, NY, USA: Nova Science Publishers, Inc.

Schippa, S., Conte, M. P. (2014). Nutrients dysbiotic events in gut microbiota: Impact on human health. Nutrients, 6 (12): 5786–5805.

Schwede, S., Thorin, E., Lindmark, J., Klintenberg, P., Jääskeläinen, A., Suhonen, A., Laatikainen, R., Hakalehto, E. (2017). Using slaughterhouse waste in a biochemical based biorefinery - results from pilot scale tests. Environmental Technology, 38: 1275–1284.

Sen, M., Lahane, A., Lahane, T. P., Parekh, R., Honavar, S. G. (2021). Mucor in a viral land: A tale of two pathogens. Indian Journal of Ophthalmology, 69 (2): 244–252.

Shin, A., Preidis, G. A., Shulman, R., Kashyap, P. C. (2019). The gut microbiome in adult and pediatric functional gastrointestinal disorders. Clinical Gastroenterology and Hepatology, 17 (2): 256–274.

Song, C. W., Park, J. M., Chung, S. C., Lee, S. Y., Song, H. (2019). Microbial production of 2,3-butanediol for industrial applications. Journal of Industrial Microbiology & Biotechnology, 46: 1583–1601.

Spaan, A. N., Van Strijp, J. A. G., Torres, V. J. (2017). Leucocidins: Staphylococcal bi-component pore-forming toxins find their receptors. Nature Reviews Microbiology, 15: 435–447.

Steiger, M. G., Blumhoff, M. L., Mattanovich, D., Sauer, M. (2013). Biochemistry of microbial itaconic acid production. Front Microbiol, 4. Available from https://www.ncbi.nlm.nih.gov/pmc/articles/PMC3572532/.

Svetlitchnyi, V., Kensch, O., Falkenhan, D., Korseska, S., Lippert, N., Prinz, M., Sassi, J., Schickor, A., Curvers, S. (2013). Single-step ethanol production from lignocellulose using novel extremely thermophilic bacteria. Biotechnology for biofuels. 6. 31. 10.1186/1754-6834-6-31.

Taheri, F., Nazarian, S., Ahmadi, T. S., Gargari, S. L. M. (2020). Protective effects of egg yolk immunoglobulins (IgYs) developed against recombinant immunogens CtxB, OmpW and TcpA on infant mice infected with Vibrio cholerae. International Immunopharmacology, 89. Pt B: 107054. doi: 10.1016/j.intimp.2020.107054. Epub 2020 Oct 13. PMID: 33065385.

Usman, H. M., Abdulkadir, N., Gani, M., Maiturare, H. M. (2017). Bacterial pigments and its significance. MedCrave Publishing, 4. Available from https://medcraveonline.com/MOJBB/MOJBB-04-00073.pdf.

Van Immerseel, F., Boyen, F., Gantois, I., Timbermont, L., Bohez, L., Pasmans, F., Haesebrouck, F., Ducatelle, R. (2005). Supplementation of coated butyric acid in the feed reduces colonization and shedding of Salmonella in poultry. Poultry Science, 84 (12): 1851–1856. doi: 10.1093/ps/84.12.1851. PMID: 16479940.

Vanhoutte, I., Audenaert, K., De Gelder, L. (2016). Biodegradation of mycotoxins: Tales from known and unexplored worlds. Frontiers in Microbiology, 7: 561.

Villéger, R., Lopès, A., Carrier, G., Veziant, J., Billard, E., Barnich, N., Gagnière, J., Vazeille, E., Bonnet, M. (2019). Intestinal microbiota: A novel target to improve anti-tumor treatment? International Journal of Molecular Sciences, 20 (18): 4584. doi: 10.3390/ijms20184584. PMID: 31533218; PMCID: PMC6770123.

Wasieleski, J. (2018). Fermented Ingredients for Natural Preservation. Kerry Health and Nutrition Institute. Available from: https://khni.kerry.com/news/blog/fermented-ingredients-for-natural-preservation/

Weizmann, C. (1949). Trial and Error, The Autobiography of Chaim Weizmann. Philadelphia, USA: The Jewish Publication Society of America.

Wemmenhove, E., Van Valenberg, H. J. F., Zwietering, M. H., Van Hooijdonk, T. C. M., Wells-Bennik, M. H. J. (2016). Minimal inhibitory concentrations of undissociated lactic, acetic, citric and propionic acid for Listeria monocytogenes under conditions relevant to cheese. Food Microbiology, 58: 63–67.

Xu, Z., Jiang, L. (2019). Butyric acid. In: Moo-Young, M. (ed.). Comprehensive Biotechnology. Elsevier Science & Technology.

Xu, Z., Shi, Z., Jiang, L. (2019). Acetic and propionic acids. In: Moo-Young, M. (ed.). Comprehensive Biotechnology. Elsevier Science & Technology.

Xue, C., Zhao, J., Chen, L., Yang, S. T., Bai, F. (2017). Recent advances and state-of-the-art strategies in strain and process engineering for biobutanol production by Clostridium acetobutylicum. Biotechnology Advances, 35 (2): 310–322. doi: 10.1016/j.biotechadv.2017.01.007. Epub 2017 Feb 3. PMID: 28163194.

Yuille, S., Reichardt, N., Panda, S., Dunbar, H., Mulder, I. E. (2018). Human gut bacteria as potent class I histone deacetylase inhibitors in vitro through production of butyric acid and valeric acid. PLoS ONE. Public Library of Science, 13: e0201073.

Mikko Immonen, Ndegwa H. Maina, Rossana Coda, Kati Katina

5 Upcycling of surplus bread using tailored biotransformation

Abstract: Edible surplus bread from the baking industry is mostly downgraded for bioethanol production or animal feed. However, novel biotransformation processes provide opportunities for safe and efficient upcycling of the surplus bread within bakery production. Tailored fermentation (sourdough technology) and enzymatic treatments can transform the technological and nutritional features of the surplus bread to better fit for recycling as a new dough ingredient. Lactic acid bacteria can be used to acidify the surplus bread matrix and to produce texture-enhancing exopolysaccharides or antifungal peptides. Furthermore, bakery enzymes can be used to degrade gelatinized starch or denatured gluten in the surplus bread to obtain higher amounts of fermentable sugars or free amino nitrogen. In this chapter, the recent progress and perspectives of surplus bread biotransformation are being reviewed, aiming at complete utilization of the material.

5.1 Generation of surplus bread and challenges of its recycling

Enhanced resource efficiency is one of the key approaches toward more sustainable food systems that are being demanded by increasing population and environmental challenges. Side streams and excess food can no longer be wasted, but should be utilized for human consumption when they retain food-grade quality (Otles et al., 2015). Cereal grains compose more than half of all wasted food calories (Gustavsson et al., 2011). Nutritious outer grain layers are often discarded in flour milling, but a huge amount of surplus bakery products are generated or spoiled in bakeries, retail, and households. It has been estimated that in Finnish bakeries 5–10% of the total bakery production volume ends up wasted. This mostly edible surplus material is currently

---

Mikko Immonen, Department of Food and Nutrition, University of Helsinki, Helsinki, Finland,
e-mail: mikko.o.immonen@helsinki.fi
Ndegwa H. Maina, Department of Food and Nutrition, University of Helsinki, Helsinki, Finland,
e-mail: henry.maina@helsinki.fi
Kati Katina, Department of Food and Nutrition, University of Helsinki, Helsinki, Finland,
e-mail: kati.katina@helsinki.fi
Helsinki Institute of Sustainability Science, University of Helsinki, Finland
Rossana Coda, Department of Food and Nutrition, University of Helsinki, Helsinki, Finland,
e-mail: rossana.coda@helsinki.fi

https://doi.org/10.1515/9783110724967-006

used for bioethanol production, animal feed, or composting, none of which are economically nor environmentally satisfactory means. Surplus bread is the single largest edible and under-utilized bakery side stream. Upcycling the surplus bread as a new baking ingredient would be an optimal utilization, however, challenges related to safety, feasibility, and technological aspects of the process have to be resolved.

Recycling bread as such has already proven to be detrimental to the volume and texture of bread (Immonen et al., 2020; Goshima et al., 2019). It has been proposed that the gelatinized starch in surplus bread hinders gluten network formation by withdrawing available water, thus preventing optimal gluten hydration and by disrupting the gluten continuous phase. Other factors such as dilution of vital gluten and native starch might also negatively alter the dough development and subsequent bread quality. In addition, activation of *Bacillus* spp. spores that are occasionally present in baking ingredients and might survive baking conditions (De Bellis et al., 2015) must be prevented during the recycling. Toxin-producing *B. cereus* and rope-forming *B. subtilis*, for instance, raise a direct hygienic concern (De Bellis et al., 2015; Katina et al., 2002). High water activity, temperature above 25 °C and pH above 5 are considered as favorable conditions for spore activation (Valerio et al., 2012) and, therefore, need to be controlled in order to maintain a safe process.

5.2 Upcycling through biotransformation

Wheat flour is by far the most common bread raw material worldwide, mostly due to the wheat dough viscoelastic gluten matrix capable of entrapping gas and producing light, fluffy, and soft breadcrumb favored by consumers. Wheat is superior from the technological perspective of bread production, as other grains fail to produce a viscoelastic dough matrix. Cereal products containing upcycled ingredients tend to be preferred by modern consumers indicating that resource efficiency is increasingly being valued on the market (Grasso and Asioli, 2020). Another present consumer trend is toward "Clean-label" food, which generally describes products with natural ingredients and no E-coded food additives. Production of such clean-label foods with high-quality texture and flavor as well as good preservation, among other measures, can be obtained with the efficient use of ingredients and processing instead of food additives. A biotransformation approach, using microbes and enzymes, provides a spectrum of alternative processing means to enhance healthiness, technological quality, and even consumer acceptance of foods, enabling the upcycling of products such as surplus bread as presented in Figure 5.1. The benefits of biotransformation are still largely unexplored.

Microbial fermentation including lactic acid bacteria (LAB) and yeasts, for example, is one of the key approaches to grain-based matrices biotransformation. Indeed, sourdough technology is the oldest and the most complete way of leavening

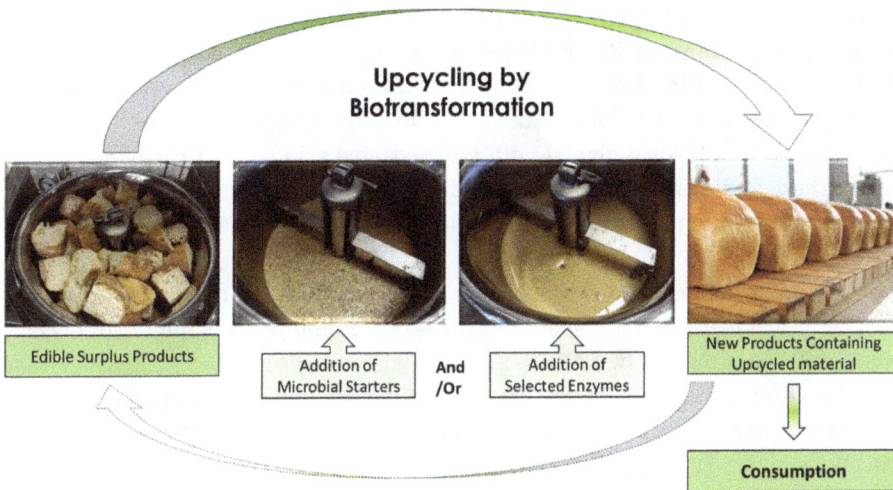

Figure 5.1: Schematic presentation of surplus bread upcycling process by biotransformation.

bread or bread-like foods (Arora et al., 2021). In the nineteenth century, baker's yeast started to replace sourdough baking, providing fast and repeatable leavening, simplifying also industrial scale baking. According to present knowledge, however, sourdough containing a naturally varying population of LAB and yeast has been proven superior compared to baker's yeast and chemical leavening (Poutanen et al., 2009; Gobbetti et al., 2014). The most well-known functional and health aspects of consuming sourdough products have been extensively reviewed (Arora et al., 2021), including, improved sensory properties, lowered glycemic index of bread, improved bread digestibility, improved mineral bioavailability, and increased vitamin content. The enhancement of bread flavor during sourdough fermentation is mostly due to the production of organic acids, activation of endogenous flour enzymes due to pH decrease and formation of various volatile aroma compounds (Arora et al., 2021). In addition to health and flavor, using selected LAB strains, it is possible to prolong bread shelf life, reduce salt content, and improve technological attributes of bread such as volume and softness (Katina et al., 2006; Rizzello et al., 2014; Arora et al., 2021). Therefore, sourdough is currently and increasingly making a large-scale comeback in domestic bread making as well as in industrial bakeries.

Surplus bread as a processing matrix is somewhat different to flour itself. During baking, the proteins are denatured and starch gelatinized, losing their original functionalities. Furthermore, enzymes present in the flour are inactivated and active microbes killed due to thorough heating of the product. These significant alterations of the matrix provide an interesting platform for biotransformation performed using selected enzymes and microbial starters. Gelatinized starch is now directly available for amylolytic enzymes. Surplus bread fermentation for new dough production was originally studied in the late 1990s (Gélinas et al., 1999). Since then, scientific research exploring bread recycling as a new bread ingredient has been non-existent until recent years. Surplus bread has been proven as a practical microbial cultivation substrate (Verni et al., 2020). A recent study revealed that the buffering capacity of surplus wheat bread matrix is very low compared to flour, which enables a fast pH drop (to pH 4) combined with low titratable acid content (Immonen et al., 2020). These findings encourage the use of lactic acid fermentation (low pH) as a hygienic control without a major risk of impaired sensory perception caused by intense acidification.

Specific selection of LAB for surplus bread fermentation can further upgrade the functional features of the material, therefore, enabling upcycling. Exopolysaccharides (EPS) produced by LAB are natural hydrocolloids that can be beneficial for bread texture. Especially linear, high-molecular-weight dextran-type EPSs have proven effective in retarding staling and increasing loaf volume in various bread applications, when produced in situ by fermentation (Katina et al., 2009; Wang et al., 2019). The same quality-improving effect was recently observed when surplus bread was fermented with dextran-producing *Weissella confusa* A16 strain (Immonen et al., 2020). In the experiment, 12.5% surplus bread was used to replace flour in the bread dough. Fermenting the bread with *W. confusa* together with sucrose (substrate for dextran synthesis) allowed a bread volume increase of over 10% and a crumb hardness decrease of almost 40% compared to non-fermented breads. Dextran production is often accompanied with isomalto-oligosaccharide synthesis due to presence of acceptor molecules such as maltose (Leemhuis et al., 2013). These isomalto-oligosaccharides have prebiotic properties, which brings another health aspect to fermenting with dextran-producing strains, in addition to technological improvements. Considering the wide range of structurally versatile EPS produced by LAB (Galle and Arendt, 2014), many bakery applications beyond dextrans are to be expected when research in the field progresses.

Another important functional feature of using select LAB for sourdough fermentation, is the production of antifungal ingredients which provides a clean-label extension of bread mold-free time (Arora et al., 2021). A wide range of carboxylic acids and antifungal peptides produced by LAB provide a possibility to tailor fermentations for product preservation needs. An example of producing antifungal bread ingredients by fermentation of surplus bread was successfully performed by Nionelli et al. (2020). The significant antifungal effect of *Levilactobacillus brevis* AM7 (formerly *Lactobacillus brevis*) was observed in a pilot bakery environment, where breads containing fermented surplus bread hydrolysate became moldy several days after the respective control breads. The antifungal activity was due to the result of synergistic effect of bioactive peptides produced by *L. brevis* AM7 using hydrolyzed bread proteins as substrates.

Enzymes are a solid part of the modern baking industry. When considering feasible upcycling methods for surplus bread through biotransformation processes, the use of common bakery enzymes may be beneficial. Especially bread starch hydrolysis by amylolytic enzymes is of interest for two reasons. First, as discussed above, gelatinized starch in surplus bread is in its original state harmful for new bread dough development, volume, and texture. Breaking down large starch molecules reduces starch water absorption and potential hindering of the gluten network, therefore, allowing better hydration of other dough components (Goesaert et al., 2008) and development of continuous gluten, necessary for good-quality bread. Second, starch hydrolysis can be tailored to result in end products of interest, such as malto-oligosaccharides, maltose, or glucose, that can further be useful for baking purposes. Malto-oligosaccharides may act as anti-staling compounds (Miyazaki et al., 2004) and glucose can be used to replace sucrose that is often added to accelerate leavening, hence cleverly improving the feasibility of the process. In the recent study, hydrolysis of bread gelatinized starch improved the dough development, bread volume and texture (Immonen et al., 2021). Starch hydrolysis into malto-oligosaccharides and maltose retarded bread staling, however, the dilution of native gluten proteins prevented from reaching the specific volume of the control wheat bread. Production of syrups from surplus bread does not, of course, limit the upcycled material for bakery usage, but the most suitable food application should be sought. Furthermore, in some cases, it may be reasonable to combine the enzymatic treatment with fermentation. Some microbes of interest may require a substrate composition, for example, rich in fermentable sugars or free amino nitrogen, that are not present in surplus bread unless additional enzymes are involved.

5.3 Conclusions

Tailored biotransformation with selected LAB and enzymes can improve the functionality of surplus bread, therefore, enabling upcycling within bakery products. The designed processes should aim at ensuring the safety, technological quality, and complete utilization of surplus material. Upcycling of food-grade side streams is necessary for increasing resource efficiency and prevention of food waste, which both are key actions for feeding the growing population and reducing the climate impact of the food system included in the United Nations Agenda for Sustainable Development.

References

Arora, K., Ameur, H., Polo, A., Di Cagno, R., Rizzello, C. G., Gobbetti, M. (2021). Thirty years of knowledge on sourdough fermentation: A systematic review. Trends in Food Science and Technology, 108, 71–83.

de Bellis, P., Minervini, F., Di Biase, M., Valerio, F., Lavermicocca, P., Sisto, A. (2015). Toxigenic potential and heat survival of spore-forming bacteria isolated from bread and ingredients. International Journal of Food Microbiology, 197, 30–39.

Galle, S., Arendt, E. K. (2014). Exopolysaccharides from Sourdough Lactic Acid Bacteria. Critical Reviews in Food Science and Nutrition, 54, 891–901.

Gélinas, P., Mckinnon, C. M., Pelletier, M. (1999). Sourdough-type bread from waste bread crumb. Food Microbiology, 16, 37–43.

Gobbetti, M., Rizzello, C. G., Di Cagno, R., De Angelis, M. (2014). How the sourdough may affect the functional features of leavened baked goods. Food Microbiology, 37, 30–40.

Goesaert, H., Leman, P., Delcour, J. A. (2008). Model approach to starch functionality in bread making. Journal of Agricultural and Food Chemistry, 56, 6423–6431.

Goshima, D., Matsushita, K., Iwata, J., Nakamura, T., Takata, K., yamauchi, H. (2019). Improvement of bread dough supplemented with crust gel and the addition of optimal amounts of bakery enzymes. Food Science and Technology Research, 25, 625–636.

Grasso, S., Asioli, D. 2020. Consumer preferences for upcycled ingredients: A case study with biscuits. *Food Quality and Preference*, 84.

Gustavsson, J., Cederberg, C., Sonesson, U., van Otterdijk, R., Meybeck, A. (2011). Global Food Losses and Food Waste: Extent, Causes and Prevention. Rome: Food and Agriculture Organisation of the United Nations.

Immonen, M., Maina, N. H., Wang, y., coda, R., Katina, K. (2020). Waste bread recycling as a baking ingredient by tailored lactic acid fermentation. International Journal of Food Microbiology, 327.

Immonen, M., Maina, N. H., Coda, R. & Katina, K. 2021. The molecular state of gelatinized starch in surplus bread affects bread recycling potential. *LWT*, 150.

Katina, K., Heiniö, R. L., Autio, K., Poutanen, K. (2006). Optimization of sourdough process for improved sensory profile and texture of wheat bread. LWT – Food Science and Technology, 39, 1189–1202.

Katina, K., Maina, N. H., Juvonen, R., Flander, L., Johansson, L., Virkki, L., Tenkanen, M., Laitila, A. (2009). In situ production and analysis of Weissella confusa dextran in wheat sourdough. Food Microbiology, 26, 734–743.

Katina, K., Sauri, M., Alakomi, H. L., Mattila-Sandholm, T. (2002). Potential of lactic acid bacteria to inhibit rope spoilage in wheat sourdough bread. LWT – Food Science and Technology, 35, 38–45.

Leemhuis, H., Pijning, T., Dobruchowska, J. M., van Leeuwen, S. S., Kralj, S., Dijkstra, B. W., Dijkhuizen, L. (2013). Glucansucrases: Three-dimensional structures, reactions, mechanism, α-glucan analysis and their implications in biotechnology and food applications. Journal of Biotechnology, 163, 250–272.

Miyazaki, M., Maeda, T., Morita, N. (2004). Effect of various dextrin substitutions for wheat flour on dough properties and bread qualities. Food Research International, 37, 59–65.

Nionelli, L., Wang, Y., Pontonio, E., Immonen, M., Rizzello, C. G., Maina, H. N., Katina, K., Coda, R. (2020). Antifungal effect of bioprocessed surplus bread as ingredient for bread-making: Identification of active compounds and impact on shelf-life. Food Control, 118.

Otles, S., Despoudi, S., Bucatariu, C., Kartal, C. (2015). Food waste management, valorization, and sustainability in the food industry. Food Waste Recovery: Processing Technologies and Industrial Techniques.

Poutanen, K., Flander, L., Katina, K. (2009). Sourdough and cereal fermentation in a nutritional perspective. Food Microbiology, 26, 693–699.

Rizzello, C. G., Calasso, M., Campanella, D., de Angelis, M., Gobbetti, M. (2014). Use of sourdough fermentation and mixture of wheat, chickpea, lentil and bean flours for enhancing the nutritional, texture and sensory characteristics of white bread. International Journal of Food Microbiology, 180, 78–87.

Valerio, F., de Bellis, P., Di Biase, M., Lonigro, S. L., Giussani, B., Visconti, A., Lavermicocca, P., Sisto, A. (2012). Diversity of spore-forming bacteria and identification of Bacillus amyloliquefaciens as a species frequently associated with the ropy spoilage of bread. International Journal of Food Microbiology, 156, 278–285.

Verni, M., Minisci, A., Convertino, S., Nionelli, L., Rizzello, C. G. (2020). Wasted Bread as Substrate for the Cultivation of Starters for the Food Industry. Frontiers in Microbiology, 11.

Wang, Y., Compaoré-Sérémé, D., Sawadogo-Lingani, H., Coda, R., Katina, K., Maina, N. H. (2019). Influence of dextran synthesized in situ on the rheological, technological and nutritional properties of whole grain pearl millet bread. Food Chemistry, 285, 221–230.

Elias Hakalehto, Anneli Heitto, Frank Adusei-Mensah, Alli Pesola,
Jouni Pesola, Robert Armon

6 Different strategies for viral and bacterial prevention and eradication from foods

Abstract: Essential features of healthy, safe, and nutritive foods are freshness and cleanliness. These characteristics need to be evaluated both microbiologically and chemically. An important prerequisite is the high quality of the raw materials as well as the production process.

When the food enters our digestive system, it brings along the outside microflora. Many fermented foods have been produced with special seed cultures, in specific conditions. All the external microorganisms meet our body defenses, such as low stomach pH, bile secretions, immune system, slime, and defensive peptides.

Moreover, the incoming bacteria, viruses, and other microorganisms encounter the human normal flora or microbiome.

Hygiene monitoring means the keeping up of food or drinks, which are free from pathogens or otherwise harmful organisms or chemicals, toxins, infective proteins (prions), allergenic molecules, and unhealthy components. The two last factors are more or less subjective. Harmful microbes include antibiotic-resistant strains or strains that otherwise disturb the balance of the alimentary microbiome.

In order to monitor these important criteria of healthy foods, many laws and regulations have been established. The awareness and knowledge of the customers about food security need to be increased. Several microbiological technologies and approaches are helping food professionals in evaluating and estimating the quality of foods.

6.1 Introduction

Hazards of pandemic dissemination are not just separated threats or unpleasant scenarios for the society or its individuals. They bring about an increase in adverse

Elias Hakalehto, Finnoflag Oy, Kuopio and Siilinjärvi, Finland; Department of Agricultural Sciences, University of Helsinki, Helsinki, Finland; University of Eastern Finland, Kuopio, Finland
Anneli Heitto, Finnoflag Oy, Kuopio and Siilinjärvi, Finland
Frank Adusei-Mensah, Finnoflag Oy, Kuopio and Siilinjärvi, Finland; University of Eastern Finland, Kuopio, Finland
Alli Pesola, Finnoflag Oy, Kuopio and Siilinjärvi, Finland; Institute of Biomedicine, University of Eastern Finland, Kuopio, Finland
Jouni Pesola, Department of Paediatrics, Kuopio University Hospital, Kuopio, Finland; University of Eastern Finland, Kuopio, Finland
Robert Armon, Technion, Israel Institute of Technology, Haifa, Israel

https://doi.org/10.1515/9783110724967-007

health effects on the human community, in general. On the individual level, an infection is ruining or compromising our health in many ways. In the long term, the remission may be followed by many repercussions and sensitivities, which have also been called the "long Covid."

Correspondingly, society is paralyzed concerning many of its essential functions. We could well say that the microbiological imbalance or the hygiene impairment or the loss of functions is a disaster for the society at all levels. It is like a knife cutting through all the layers of a cake. Such effects cannot be measured by counting the economic losses only. And we should be prepared for the dire consequences in the future, at the same time as we are getting ready to encounter possible future pandemics. Therefore, the right timing of preventive measures is of crucial importance.

Unpreparedness in facing the outside threats is directly proportional with the time required for the recovery of society to get out of the pandemic situation, or at least to return to a level where an upper hand has been reached over the contagious agents and their aftermath. One has to get alerted with and prepared for the shortages and cuts in the resources and storage, or deficiencies in the safety of the distribution chains. This book is intended for making a platform for discussion that could make a difference in ordinary people's lives, that is, us. The correct layout of the food production, storage and transportation systems is precious.

Plentiful options are available, but the information, knowledge, and experience must be at hand when needed. The level of the infection, its infectiveness, and the current epidemics situation are important to be acknowledged. Also, the variants and complications or aftermath of the actual disease need to be taken into account early enough. For instance, as Professor Jeremy K. Nicholson stated in his lecture at Virtual Metabolomics Symposium on June 22, 2021, patients with post-acute COVID-19 had a 2–10 times elevated risk for getting myocardial infections. The title of his lecture was "Evaluation of Post-Acute COVID-19 Syndrome using NMR Spectroscopy and Mass Spectrometry" (Holmes et al., 2021).

We want to support the crucial discussion among those who take care of the vital services in society. Safe food distribution is at stake. Consequently, there is a need to remain in good health and shape. One fast preventive procedure against emerging new pandemics could be the passive immunization with chicken egg yolk antibody (Hakalehto, 2021a). To develop the details and see the big picture warrants in infection protection, exchange of ideas and teamwork. It is in the hands of those professionals who investigate the microbes, but also with those who take care of the cold chain, shelf life, distribution, client surface (encounter with the customers), and so on. We also need to establish a constant search for novel raw materials, methods, techniques, and recipes, which means, in a way reinventing food.

6.1.1 History of food spoilage

The different critical points where the food materials could get contaminated should be carefully investigated and identified. This is equally true with the acquisition, formulation, transport, and storage of the raw materials, as it is for the actual production process and the products themselves.

Different food materials get spoiled in different ways and by partially distinct microorganisms. In his book *Modern Food Microbiology* (1986), James M. Jay, PhD, is making a fundamental survey on the issue of the microbes in food. He is the former chairman of the Food Microbiology Division of the American Society of Microbiology. Since the publication of his book, the microbiological monitoring of foods has somewhat changed as genetic techniques and other new methods, have developed. They have been implemented into industrial hygiene control. However, cultivation techniques still form the baseline in food microbiology. It is noteworthy, though, that food spoilage or the contagion via food has to be seen in a wider context than only as an intrusion of a contaminant or a pathogen into the food or its components or compartments of the food in the body system. This requires a reorientation of the microbiome balances and the balances with the host. It is always an encounter or challenge by the outer environment microbes, microbes of the existing microbiome of the food, or in the body system. Moreover, it is a matter of unit microbial communities often forming a network or microbiological exchange of signals.

This exchange of microflora takes place in the families, where adults, children, and domestic animals or plants share many microbial strains. In fact, the "intruder" strain may often be a member of the so-called normal flora becoming active or harmful due to some change in the conditions.

Dr. Jay presents in his excellent book on more than four pages the activities, or milestones, in combatting microbial contaminations in foods. This includes 106 individual events throughout three and a half-century. Here, we list a few most important ones according to the scope of the current book:

1659 Kirchner demonstrated bacteria in milk.
1782 Canning of vinegar was introduced in Sweden.
1825 US patent for preserving food in tin cans (T. Kensett and E. Daggett).
1835 A patent was granted to Newton in England for making condensed milk.
1857 Pasteur showed that the souring of milk was caused by microbial growth.
1865 Artificial freezing of fish started commercially in the United States
1873 Lister isolated *Streptococcus lactis* in pure culture.
1874 Sea transportation of iced meat at sea had begun.
1876 Tyndall observed the bacterial decomposition to be always traceable by chemical substances to air or the containers.
1878 First successful cargo of frozen meat from Australia to England.
1878 Cienkowski isolated *Leuconostoc mesenteroides* from sugar slimes.
1880 The pasteurization of milk began in Germany.
1888 Gaertner first isolated *Salmonella enteritidis* as a causative agent of food poisoning.

1889	The artificial freezing of eggs.
1896	Van Ermengen discovered *Clostridium botulinum*.
1902	The term psychrophilic was invented by Schmidt-Nielsen for microorganisms that grow at 0 °C.
1907	E. Metchnikoff and co-workers isolated *Lactobacillus bulgaricus* from yoghurt.
1908	Sodium benzoate permitted as a preservative in certain foods in the United States
1915	*Bacillus coagulans* isolated from coagulated milk by B. W. Hammer.
1917	*Bacillus stearothermophilus* isolated from cream-style corn by P. J. Donk.
1929	A patent in France was issued for the use of high-energy radiation for the food processing.
1938	Outbreaks of *Campylobacter enteritis* traced to milk in Illinois, USA.
1945	Food poisoning by *Clostridium perfringens* (*welchii*) first diagnosed by McClung.
1955	Similarities between cholera and *Escherichia coli* gastroenteritis in infants first demonstrated by S. Thompson.
1963	The salmonellae surveillance program in the United States.
1971	First US food-borne outbreak of *Vibrio parahaemolyticus* gastroenteritis occurred in Maryland.
1975	*Salmonella* enterotoxin demonstrated by L. R. Koupal and R. H. Deibel.
1978	Food-borne gastroenteritis caused by the Norwalk virus documented in Australia.
1979	Food-borne gastroenteritis caused by non-01 *Vibrio cholerae* occurred in Florida (Earlier outbreaks in Czechoslovakia, Europe, 1965, and Australia, 1973)
1983	*Campylobacter jejuni* enterotoxin described by Ruiz-Palacios et al. (1983).

6.2 Systematic avoidance of contamination

Spoilage is not the only way food gets contaminated. Hazardous germs may just be transferred into food, or they could use food as a vector for reaching new niches or opportunities. If these agents are cellular microbes, viruses, or prions, they may occupy the food, harbor it or cover the surfaces, or hide inside it. Therefore, all the various means for the physical contact of the pathogen need to be evaluated and taken into account. The initiation of a biofilm structure out of the swarmer cells is a process that have been studied and illustrated by micrographs (Hakalehto, 2015a).

The hygiene surveillance also requires the closing out of the initial infection or spoilage of the materials. In these precautionary measures, cold chains play a central role. This also includes distributing foods or meals to the households or other clients, such as day care, schools, and elderly homes. For example, fresh fish is easily spoiled, and it can also cause risky infections. Any salads and mayonnaise-containing foods or meals also require extra caution.

International standards and norms, and practices in food handling, such as Hazard Analysis and Critical Control Points (HACCP) could give a baseline for hygiene maintenance (Lucia Rocha Carvalho et al., 2000; Friedhoff et al., 2005). However, the epidemic situation is causing additional risks. The HACCP is originally developed besides space trips for the practices in the Texas meat industries (Carr et al., 1998).

It is nowadays a widely applicable strategies platform for dealing with pandemic hazards in food industries and healthcare.

Besides the checking points related to the cold chain or to the manufacturing process, we have to consider the encounter of the food taxis and other delivery personnel with the end-point customer. There the responsibility regarding the food quality is transferred to the client (or the next person in the delivery chain). Consequently, the entire food delivery succession has to be a secured one. For example, the food portions have to be directly handed over to the person who is the next to be responsible for them. Alternatively, the food can be placed into refrigerators or cold rooms, or in case of the hot chain, into kitchen with adequate facilities to keep the temperature. But in these various food arrival issues, the responsibility always has to follow the delivery. The food or meals should not be neglected, or left out of control.

Any education for the field personnel in the food distribution chain or working with the catering services should consist of the following:

1. keeping the food portions free from any outside contamination
2. observing the microbiological and chemical quality of food during transportation and storage
3. maintaining the cold chain (or hot chain) during deliveries
4. safeguarding the health of the personnel, protecting them from any contagious diseases
5. securing the encounter with the end-point client
6. accepting the responsibility of the transportation and the courier services, and transferring this responsibility to the next person in chain

The microbiological education should include teachings about the potential contaminants of various food varieties, ingredients, or raw materials (Hakalehto, 2015b). Microbiological testing and research are important in accumulating experience and in signposting the sequence of preemptive measures in contamination control. The concepts of indicator bacteria, as well as the basic thinking about the routes of contamination should be taught to the food-chain professionals during their education and training. During the pandemic era, it is essential to give information about the current epidemic situation, rules set by the authorities, as well as hints or methods about the protection of the customer as well as the personnel. In case of the distribution of pathogens, the environmental routes of contamination need to be evaluated (Hakalehto, 2015c).

The overall hygienic preparations warrant right attitude, correct information and recommendations, rules and orders. In practice, the alertness could also spring out from an inventive mindset of the personnel – most importantly, innovations regarding the detecting or blocking of contaminations or dissemination of infections. The preparedness needs to include an early warning system of the potential emergence of various risks. In Finland, the National Institute of Health and Welfare developed a system and techniques for monitoring viruses (SARS-CoV-2) from the wastewater (Hokajärvi et al., 2021). This approach could give real-time info about the epidemic situation in any particular community, city, or area. One example of a map from the "situation room" is shown in Figure 6.1.

Figure 6.1: Road traffic control room in Traffic Management Center, Tampere, Finland. This kind of "situation room" could be used also for the monitoring of the pandemic or epidemic prevalence of infection agents or cases. Correspondingly, for example, the microbiological cleanliness in the different parts of the distribution network could be surveilled in real time. Photo: Eero Sauramäki.

6.3 Special traits of viruses

Viruses are widely present in the ambient air. Some 100 million virus particles land on a square meter daily. These particles belong to all kinds of viruses, specific to various multicellular organisms, such as plants, animals, and humans. Also, bacterial viruses (bacteriophages) are present practically everywhere. In a laboratory, search for bacteriophages specific to any bacterial strain or species, sewage water is often used to source viral particles. Although bacterial or other microbial cells could get inactivated in the atmosphere due to dehydration or exposure to UV light or other disinfecting factors, viruses may get liberated after the lysis of bacterial or other cells. Furthermore, they may remain infective in the air because of the protection offered by the aerosol or dust particles. Viruses may stay alive within the host cells or in the microscopic particles, but they are sensitive and subject to the same denaturing conditions as bacterial cells. In fact, the viral nucleic acids are usually in a still more vulnerable position than the DNA in the much larger bacterial or eukaryotic cells.

A thumb rule for the survival or perseverance of the microbes in the microscopic airborne particles in the atmosphere: any particle which is less than 10 μm in diameter may get transported in the air relatively freely and independently of the gravitation. This makes microbes, or the particles or droplets containing them, a kind of "transferable cosmopolitans." Thus, dust from the Saharan desert sandstorm has been crossing the Atlantic Ocean in a storm that also contained many soil microbes. This dust eventually landed in the Americas (Geggel, 2020).

Correspondingly at the height of 1 km, as we sampled the atmospheric layers by a jet plane, there were at least as many mold spores as found on the meter above the soil surface (Alder-Rangel et al., 2018). And these atmospheric fungal spores are rather resistant to UV light or dehydration, or extreme temperatures.

In the big picture, this movability of various microbes often means their potential presence in surprising places or locations. Thus, the spread of viruses could occur in conditions or situations which are not easily predictable by the laboratory experiments only. As the pandemic situation has forced everyone to think more deeply about the microbial distribution, routes of contamination, microbial survival, surveillance, contagion, sedimentation, attachment, and other essential traits and characteristics of the contagious agents, we have also learned that it is not a simple issue to make predictions or calculate patterns for the microbial behavior in variable conditions. It is also noteworthy that the microbes' microscopic scale makes it difficult for us to comprehend their ways. For example, in a spoonful of yoghurt, there are as many bacterial cells as human beings on the Earth.

With this enormous space in the microcosm that is filled with microbial communities, their activities make it really hard to present assumptions on the behavior of free-moving air-borne microbes. It is the metabolic understanding, which often could explain the best microbial activities or their occurrence. However, viruses are not actively metabolizing. For example, microbial entities often occur in groups of cells or virus particles or as agglutinated cell populations or pellets of mycelia. In any such concentrated population or metabolizing cell group, there also exists a potential source of the distribution of viruses (Quammen, 2014). Thus, these densely packed biomes also offer potential sites for recombination, and in suitable physico-chemical conditions, an elevated likelihood of mutations in the very biological material in question. For example, Brazilian free-tailed bats in a cave in Carlsbad Caverns National Park in California sleep as groups of 300 individuals hanging on a wall square meter. The total number of bats in this cave could exceed a million during migration times (Hristov et al., 2010). This may further explain the generation of virus variants of SARS-CoV-2 in some of these "densely packed" populations. The wide distribution of viruses creates options for new virus variants to appear. When the COVID-19 or any other pandemics has spread globally, the risks of the new emerging variants and pathogens are imminent.

6.4 Some historical and current aspects of the coronavirus pandemics

The coronaviruses were discovered for the first time more than a century ago (Williams, 2020). They were found in various animal species. However, the first human infections caused by the coronaviruses occurred in the 1960s, and they were mild viral infections only. The first more hazardous coronavirus disease was the SARS-1, which nearly developed into pandemics in 2002–2003 (Petrosillo et al., 2020).

Similarly, the Middle East respiratory syndrome (MERS) epidemic occurred mostly in the Middle East in 2012 with a concrete threat of the viral disease of developing into global pandemics (de Wit et al., 2016). The global death percentages of SARS-1 and MERS were 10% and 34%, respectively (Wong et al., 2021).

The origin of various pandemics is an interesting scientific and pragmatic healthcare issue. The influenza epidemic ("Spanish flu"), which started in the year 1918, left several unanswered questions behind (Oxford and Gill, 2018). For example, this pandemic in 1919 was the first time when the Australian state of Queensland closed its borders, and the second time was during the COVID-19 pandemics. The pandemic episode which occurred roughly a century ago was also the first relatively well-recorded pandemic episode worldwide. According to the official statistics, it killed about 50 million people globally, but some experts suspect that the real death toll could have been twice as high. The horrible corollaries of the pandemics were worsened by much medical personnel being preoccupied with the ongoing World War I. One major reason for the spread of these past pandemics among young adults could have been the fierce "cytokine storm," which was shown to occur in the lungs of the deceased.

In fact, it is the case also with the current pandemics that the reaction of the human immune system with the pathogen largely determines the nature of the disease in individual patient cases. Although there had been numerous experiences on previous pandemic episodes globally, it seemed difficult for the governments or the global organizations to make fast enough decisions on such measures as lockdown, prevention strategies or other issues (Dasgupta and Crunkhorn, 2020). One reason is the sudden outcome of a global disease actually spreading or beginning to spread inevitably. As stated by Dasgupta and Crunkhorn (2020): "Throughout history, pandemics are integral parts of human civilization regardless of how much we progress to beat them."

Nevertheless, it is crucial to immediately start the correct precautionary preparations after recognizing the threat during the onset of the pandemic waves. Moreover, we need to understand the ways our human system reacts with the pathogens in question. Dr. Natalia Gavrilova, who designs research on the post-COVID symptoms in St. Petersburg, Russia, lectured in the Sheba Medical Center webinar on May 28, 2021. The webinar was arranged and organized by the Israeli Academician Professor Yehuda Shoenfeld in Shoenfeld's Mosaic of Autoimmunity series. In her presentation, Dr. Gavrilova listed the different symptoms associated with the post-infectious autoimmune syndrome as follows:

– Guillain–Barré syndrome (GBS) (the body system attacks its own peripheral nerves)
– Encephalomyelitis (inflammation of brain and spinal cord)
– Brainstem encephalitis
– Necrotizing autonomous syndrome

Actually, the severity of the above-mentioned post-COVID symptoms indicates the seriousness of the disease. It is of high urgency to redirect the resources toward the management of these autoimmune consequences of infections. Such health problems may occur after the initial remission in many tissues: blood circulation, kidneys, lungs, reproductive organs, liver, and digestion. The core of the enterohepatic circulation of bile substances is the meeting point of the blood circulation with the ducts of the liver leading the fluids to the bile bladder, and further to the duodenum (via the *papilla vateri* opening) (Figure 6.2). See also Hakalehto (2021b,c) and Hakalehto et al. (2019a, 2019b, 2021).

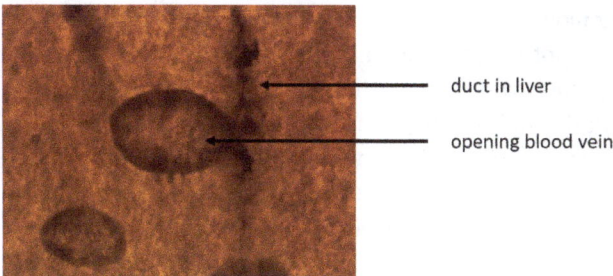

duct in liver

opening blood vein

Figure 6.2: Thin-section micrograph of calf liver (enlargement 1,000× with Nikon Eclipse E3 Microscope). This is the key area for the important enterohepatic circulation also in the human body. Here the recycled bile substances collected from the intestines are returned back to the digestive system. Photo: Elias Hakalehto in the Finnoflag Oy's laboratory.

To understand the pandemics, we have to gain knowledge and information on the causative agent, which is a virus. The reservoir of the SARS-CoV-2 virus is in the bats, according to a Wuhan–Italy research group (Platto et al., 2020). There have also risen suspects that this agent could be a reflection of the human social life (de Chadarevian and Raffaetà, 2021). There is a rather broad consensus that the current pandemic has changed future medical and basic scientific research (Mukherjee, 2020).

6.5 Case *Pectinatus* – an example of strictly anaerobic contaminant bacteria

One example of a random, but after all, most logical conditions for the contamination of food and beverages was offered in the year 1995 when we studied the occurrence of a relatively newly discovered beer-contaminating bacterial genus *Pectinatus* in real-life circumstances (Hakalehto, 2000). This bacterium was first described by Lee et al. (1978), but it was soon isolated also from many other countries such as Germany (Schleifer et al., 1990) and Finland (Haikara et al., 1981).

The distribution of *Pectinatus* contamination in breweries is somewhat parallel to the spread of the coronaviruses in the air. This case description given below also provides us an idea about the spread of viral, bacterial, or other microbial contaminants during the industrial food or drink manufacturing processes.

The Latin name *Pectinatus* means "comb-like", which refers to the flagellation of these bacterial cells on only one side of the cells (Lee et al., 1981). These organs for the cellular movement occurring only on the other side of the cell in this bacterium make it look comb-like (the flagellae, of course, are long and curvy organelles, but nevertheless, they protrude out of the cells). The bacterial cell is able to detach them, and they grow anew in 3 sec only (McNab, 2004; Hakalehto, 2000). The flagellae are rotating around their axis either clockwise or counterclockwise. The choice of the direction of this movement determines the ultimate swarming course of the bacterium in liquid, where its sophisticated and effective but simple molecular sensing system associated with the "flagellar motor" moves the cell into favorable direction.

Naturally, the viruses do not possess such a system for their movement; instead, they are moved along the currents, agglutinating or sedimenting molecular forces in different liquids. In the ambient air, viruses and bacteria are often inside tiny droplets of liquid or aerosol particles. It is these particles then, whose movements in the air determine the risk of contamination at different distances from the source of contamination. Air currents usually have a decisive impact on the movement of these particles. It is important to keep in mind that an individual aerosol, dust, or other particle or droplet may contain tens if not thousands or more of infective units. Therefore, instead of counting the number of cells or viruses within a certain volume of air or liquid, it is advisable to speak about colony forming units (CFUs) or plaque-forming units (PFUs). CFU refers to the growth of microbial colonies on a Petri dish. PFU means an empty dot or plaque on a fully grown bacterial carpet on a Petri dish or other cell culture on an agar medium, which indicates the presence and refers to the concentration of the viruses in the chosen volume of the inoculum.

In the case of the *Pectinatus* sp. bacteria in the air, one clearly observable thing is the scale difference to the regular size of objects more familiar to our senses. Namely, within the invisible liquid particles, the conditions can be favorable for the strictly anaerobic or anoxic *Pectinatus* cells. In other words, this bacterium does not tolerate any oxygen, but it still can get preserved in the air and move with it long distances within the aerosols. This example teaches us to the importance of the microscopic dimensions, niches, and conditions in them with respect to the spread of microbes as well as virus particles. For example, some circumstances or factors in the environment or in the surrounding medium or air may essentially protect the germs often in an unpredictable fashion. The viruses could get protection in this way by the particles or matrices around them, even against disinfecting vapors, gases, or UV light. For this reason, it is always of crucial importance to monitor the effectiveness of any treatment on the microbial cells or viruses in real-life conditions.

In the brewery, the surroundings were investigated by the Rapid Microbial Detection project in Kuopio during the years 1992–95 and Finnoflag Oy in 1995 onward. There were lines of empty clean bottles and relatively empty returned bottles in the one and same big industrial hall in the opposite ends of it (Hakalehto, 2000). The dirty bottle line moved these returned bottles to the station where they were washed, cleaned, and dried for reuse. In the bottling machine, these bottles were positioned in baskets of 24 bottles each. The *Pectinatus* contaminations occurred mostly during summertime and this strictly anaerobic bacterium is a really tedious one as it converts the product into a turbid and disgusting broth smelling like rotten eggs. Such product rounds are highly uncomfortable also for the reputation of the brewery, whose logo is printed onto the etiquette or label on the side of the bottles.

By that time, an unexplained observation was that in one basket only some part of the bottles randomly contained the spoiled product, whereas the other bottles in the same basket were packed with faultless products. This observation had been made in all different breweries where this anaerobic bacterium had occurred in Finland in the non-pasteurized brewery products. Unexplainable aspects, however, turned out to actually lead to understanding the airborne distribution of this bacterial contaminant. In the big industrial hall, the bottling took place in one end of the huge space. There the bottling machine forcefully "injected" the liquid into the bottles in the basket, and in half a second each of them was then filled and closed with a tight cap. The speed of bottling machine did not leave too much time for the product to get oxidized or aerated. This protected the product from chemical oxidation quite well, but simultaneously it increased the risk of contamination by anaerobic bacteria. In unpasteurized beer, it is usually the oxygen, which causes the rapid chemical spoilage of the drink. Therefore, the restricted time frame during which the liquid was shot into the clean bottles was decisive for the chemical quality. However, this gave the opportunity for the strictly anaerobic spoilage organisms like *Pectinatus* sp. to contaminate those bottles by the airborne (in the aerosols) cells being hit by the "injected" product volumes. They took along the living bacteria from the droplets into the bottle. Thus, the anaerobic bacterial cells in the air were capable of spoiling the beer in the particular bottles they contaminated.

As the beverages often contain carbon dioxide liberating into the relatively small headspace, this eventually increased the pressure inside the bottled beer. It also further reduces the amount of oxygen. Interestingly enough, the cell wall peptidoglycan layer of *Pectinatus* sp. is exceptionally thick for a Gram-negative bacterium. Its thickness is approximately 30 nm, whereas in most other species of Gram-negative bacteria the thickness is just a few nanometers (Si et al., 2019). This peptidoglycan (murein) layer strengthens the cell's outer compartment which helps the bacterial survival in the pressurized environment of the turbid beer (Rohde, 2019). Consequently, it is relatively easy to extract in the laboratory the outer membranes of *Pectinatus* strains because of this trait by using mild hydrochloric acid (Hakalehto et al., 1984).

The occurrence of *Pectinatus* bacteria in breweries was paradoxically associated with the improved brewing technologies. This bacterium offered microbiologists a highly interesting object for their studies. The example also shows the common truth that solving one problem often creates new ones.

These brewery contaminants were mostly found in the turbid beer bottles. As Gram-negative bacteria, the *Pectinatus* cells have outer membranes, including endotoxin layers, but the species-specific characters of this layer were quite unique ones in the genus *Pectinatus* (Helander et al., 1983). For example, the endotoxin was more toxic for the tested eukaryotic cells than the lipopolysaccharides (LPS) of *Escherichia coli* (Helander et al., 1984). The different strains of the genus were typed based on the surface proteins (Hakalehto et al., 1984) and their immunoreaction (Hakalehto and Finne, 1990).

The flagellation of the cells gives *Pectinatus* sp. strains their typical motility in the anaerobic conditions where they are able to sustain. They cannot live in the presence of oxygen, but facultative anaerobic species of the genus *Salmonella* just change their mode of metabolism and survive. The *Salmonella* strains expressed more readily their attachment organelles, type 1 fimbriae, in the presence of oxygen which actually indicated their preference for oxygen in the gut (Hakalehto et al., 2007).

The basic scientific survey of the *Pectinatus* strains could give ideas about their roles and potential risks caused by industrial contaminants as like the corresponding studies on the salmonellae revealed their functions as pathogens. It is one fascinating feature in microbiology that any results of research on individual organisms could lead to further understanding about the behavior of microbes in general.

But where did the *Pectinatus* cells come from? They were actually found in different parts of the brewery complex. Within the malts or other raw materials? Or in the water? Nope, the answer is most likely the "dirty bottles," whose washing line was in the same spacious hall as the bottling unit. We made ultrasensitive measurements using a method called immune polymerase chain reaction (PCR) where the positive aspects of the sensitivity of the PCR method were combined with the practical benefits and attainable specificity of the immunoassays (Hakalehto, 2000). Both methods were in everyday use in the laboratory of Elias Hakalehto during the Rapid Microbial Detection project in 1992–95 in Kuopio, Finland. Thanks to this combination of two sophisticated techniques, we could detect even single bacterial cell in any big space if a single cell or a few cells were collected in the sample (Figure 6.3). This kind of analysis protocol requires reliable sampling and effective enrichment methods.

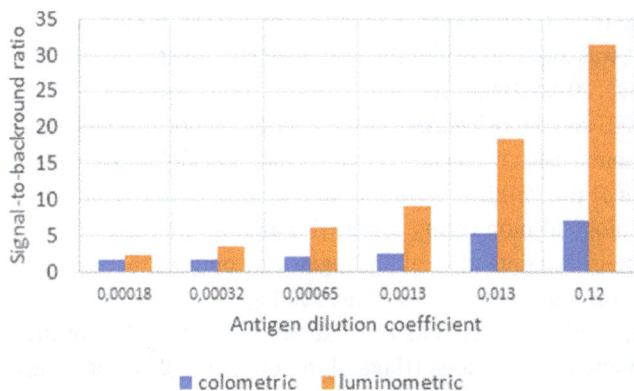

Figure 6.3: The detection of *Salmonella enterica* serovar Typhi (strain RHS 672) by using polyclonal antiserum prepared in rabbits in traditional colorimetric ELISA and in immunoluminometric immunoassays. The experimentation was carried out in the Rapid Microbial Detection (RMD) project in Kuopio, which in 1992–95 preceded the Finnoflag Oy laboratory functions. See also Kuosmanen et al. (1995) and Hakalehto et al. (2015a). Modified from Hakalehto (2000).

These technological advancements made it possible to intensify the analysis onto a previously unforeseeable level. This approach could assist in developing ultrasensitive detection of the SARS-CoV-2 viruses, too. It could be integrated with sophisticated air filtration techniques or for sieving liquids or gases. The ultrasensitive first detection of salmonellae could offer a useful reference for further method development (Hakalehto et al., 2015a).

6.6 Portable microbe enrichment unit (PMEU) in microbial detection and identification

Traditional microbiological diagnostics is based on plate culture techniques, and their variations. The rise of molecular biology has speeded up the development of faster and more sensitive methods for about 50 years. However, generally accepted standard techniques are still using the Petri dish plate cultures on solid agar media as the reference baseline. For example, we have already developed a fast and effective method for the detection of *Campylobacter* sp. (Pitkänen et al., 2009) combining the Portable Microbe Enrichment Unit (PMEU) culture with the real-time PCR (Mullis et al., 1986). This combined method of ours for the recognition of *Campylobacter* was recognized to be much faster and more effective than the international standard method (ISO 17995:2005) by the Finnish Institute of Health and Welfare in this study. It was about 10 h faster than the standard method and could produce a lot higher cell concentrations in the pre-enrichment (Hakalehto, 2010). Later, the PMEU method was developed as a field version for the detection of microaerobic *Campylobacter* sp. and *Yersinia* sp. in Burkina Faso (Hakalehto et al., 2014a). It was also applied for the fast detection of slow-growing my-cobacteria (Hakalehto, 2013b, 2015d; Hakalehto et al., 2014b, 2016a). As a field capable stand-alone system, the PMEU could provide means for the ultrafast enrichment of *Myco-bacterium tuberculosis* "in hours instead of days, in days instead of weeks" (Figure 6.4).

The different versions of the PMEU enrichment and detection device could be used for various microbiological monitoring tasks (Hakalehto and Heitto, 2012). They could assist in the sensitive detection of pathogenic or opportunistic strains which occur at low concentrations beneath other microbiological growth. For exam-ple, *Salmonella* sp. strains and their antigens were detected by the PMEU enrich-ment in both aerobic and anoxic conditions (Hakalehto et al., 2007). We have also developed ultrasensitive *Salmonella* detection based on the PMEU pre-enrichment and immunoassays or PCR techniques (Hakalehto et al., 2015a). Moreover, the inter-actions of various bacterial strains could be monitored by the PMEU in the microbio-logical specimens (Hakalehto et al., 2010; Hakalehto and Jaakkola, 2013).

Just at the time of compiling the present book, there were reports in the media that about 450 citizens in day care centers of Jyväskylä city in Finland had caught *Salmonella* infections; 75% of them being children. The cause of this incident at the end of June 2021 was most likely the imported salad for the lunches. It indicates as an example a typical *Salmonella* outbreak which often affects large groups of people via institutional catering.

6.6.1 Using PMEU in studies on the microbiome, water hygiene, antibiotic resistance, industrial monitoring

Besides the pathogenic gut organisms, we have detected the normal enterobacte-rial intestinal flora of small children using the PMEU technology (Pesola et al., 2009). The development of this important infantile subpopulation during neonatal growth was followed in another study (Pesola and Hakalehto, 2011). Then it was possible to demonstrate that the microbiome of a 3-year child was recovered in about 6 months after as many as 19 antibiotic treatments during his early years. The PMEU approach for the monitoring of antibiotic-resistant strains was reported in 2011 (Hakalehto, 2011b). The automated sampling by the automated sample collec-tion system (ASCS) is used in the PMEU versions connected with the water distribu-tion system (Hakalehto et al., 2011, 2013a, 2015b).

Additional aspects of the PMEU approach are the industrial contamination con-trol (Mentu et al., 2009), and the sensitive screening of hospital strains by the PMEU

a

b

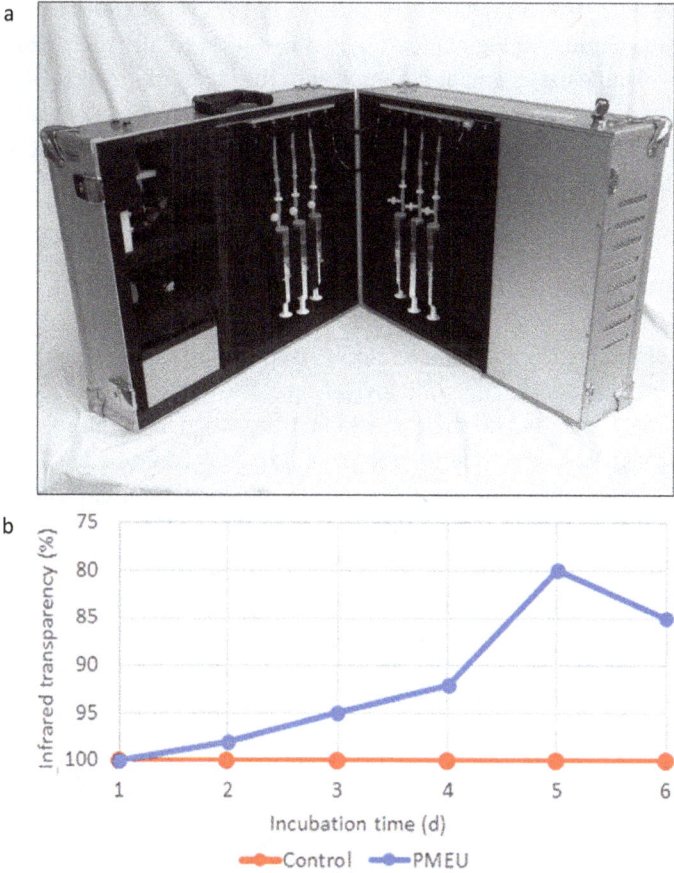

Figure 6.4: (a) Portable microbe enrichment unit (PMEU, Finnoflag Oy, Finland). (b) Growth of *Mycobacterium marinum* environmental isolate (ATCC 927) on the M7H9 broth with Middlebrook ADC enrichment supplement (Becton Dickinson, USA). Red line: growth in the control tube. Blue line: growth in the PMEU Spectrion®. Decrease in IR-transparency (y-axis) is presented here as percentages (Hakalehto et al., 2014b).

Scentrion® (Hakalehto et al., 2009; Pesola et al., 2012; Laitiomäki et al., 2015). For the hospital diagnostics, we used the industrial prototype of the PMEU, which is detecting the contaminants in a few hours of time by their gas emissions. The future potentials of the PMEU units for industrial and hospital monitoring were tested in a project where the operation theatres of two Finnish hospitals were monitored for the occurrence of antibiotic-resistant bacteria (Hakalehto et al., 2021).

The general trends with respect to the food sector have been directed toward healthier nutrition for about a century, with an accelerated scientific approach. Naturally, the often controversial question of what is the healthier food depends on the individual genes and health status of a consumer. The recent ideas and tendencies in the food sector have been discussed earlier (Hakalehto, 2020). Understanding regarding the importance of microbiological food hygiene, has originally risen from comprehending the typical features and experience with the traditional foods (Hakalehto, 2015c). In different meat or fish sources, or in the grains, fruits, or vegetables, there always can be found their specific spoilage organisms, in addition to the miscellaneous microflora.

6.6.2 PMEU in the ultrasensitive microbiological surveillance

The PMEU device by Finnoflag Oy, Finland, could be used for the rapid screening, characterization, and classification of microbiological samples. It offers tools for the distinction of bacterial groups based on their metabolic traits, such as in the case of various species of *Staphylococcus* sp. (Figure 6.5). Such methods could be used for the rapid monitoring of contaminating microbial strains in, for example, meat or dairy industries, or in the hospitals. The technical description of the PMEU approach in the food and water surveillance is presented by Hakalehto et al. (2016b). It is possible to monitor the condition of the production animals with respect to such diseases as mastitis. For example, according to the FAO, the most common mastitis agents were *Staphylococcus aureus, Streptococcus agalactiae, S. dysgalactiae, S. oberis*, and *Escherichia coli* (Hakalehto et al., 2016a). By the PMEU approach, it is possible to further specify the type of contamination in a short time. This approach was used for estimating the quality of retail milk as described by Hakalehto et al. (2013b). The PMEU technology could thus be implemented into the HACCP system. This protocol was originally developed for manned space travels but was then adapted into the production of canned foods and in the Texas meat industries (Sperber and Stier, 2009). Nowadays United States Department of Agriculture (USDA) regulates the use of HACCP in the meat industry whereas in seafood and juice production, it is regulated by the US Food and Drug Administration (FDA). Since the HACCP principle is designed also for the protection of the general public against biological threats and hazards, it could be well suitable for the monitoring of food distribution and deliveries during and after pandemics. For an example of the succession of "control points," see Figure 6.6.

Similarly, the PMEU method could be used for the speeding up of the recovery and enrichment of slow-growing mycobacteria. Actually, *Mycobacterium tuberculosis* bacterium may have caused more human deaths than all the other contagious microbial pathogens together. Inactivated *M. tuberculosis* cells are also used as adjuvants in vaccinations, for boosting the immunogenicity of the antigens. For example, such mixture is produced with the Freund´s complete adjuvant (FCA), yielding better immune responses against the desired microbial or corresponding antigen in the mixture (Freund et al., 1937). In fact, during this current COVID-19 crisis, it was anticipated that areas with a long history of compulsory Bacillus Calmette–Guérin

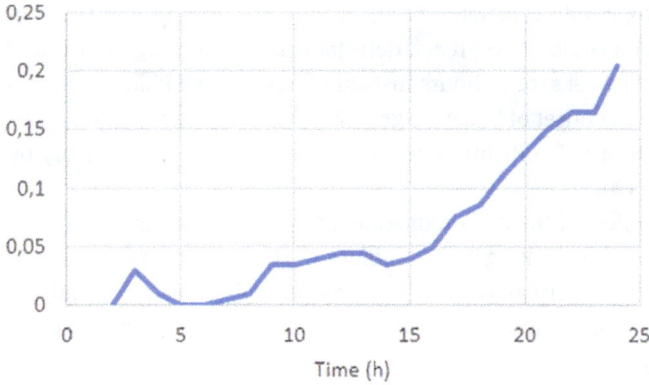

Figure 6.5: Volatiles liberated by hospital isolates of *Staphylococcus aureus* strain as measured by one of the metal oxide sensor (MOS) probes of the PMEU Scentrion®. The PMEU Scentrion® version was an industrial prototype, which also has been used for ultrafast detection of bacterial and other microbial pathogens (Hakalehto et al., 2013b; Hakalehto, 2015e). In a preliminary testing of the combined use of the PMEU and the real-time PCR, we came into conclusion that 2–4% of the mastitis bacteria could be detected with the combination of these two methods. This project was carried out in cooperation with Finnzymes Oy 7 years ago. On the Y-axis there is the relative drop in the transmission as measured by MOS. In this case, the gaseous compounds emitted on the later stages of the culture were observed, as the recovery (lag phase) of the culture with a small inoculum lasted around 8 h, and the logarithmic growth phase started after about 15 h of cultivation. Modified from Hakalehto (2012b).

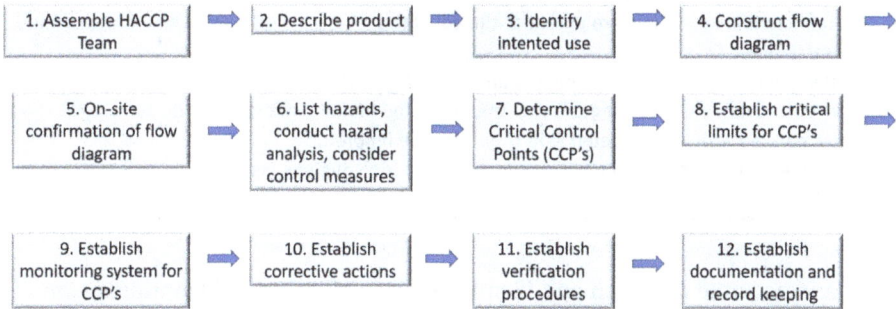

Figure 6.6: Twelve steps of HACCP process (modified, source: National Centre for HACCP Certification, India).

(BCG) vaccination coped better with the virus – such as former Eastern Germany in comparison with Western Germany, or Finland, in comparison with Sweden. However, this type of evidence has been controversial (Marzoog and Vlasova, 2021).

Our laboratory at Finnoflag Oy has speeded up the detection of the slow-growing environmental mycobacteria, such as *Mycobacterium marinum*, in the PMEU device. The results of these studies were published in the *Pathophysiology* journal in

2013 and 2014 (Hakalehto, 2013b; Hakalehto et al., 2014b). The microbial cultures of slowly propagating bacteria could at best reach detectable levels, starting from singular cells, in days instead of weeks, or in hours instead of days. In the PMEU unit, it is also possible to speed up the onset of bacterial growth by leading aseptically the carbon dioxide from one culture of cells into another one, as described previously by Hakalehto and Hänninen (2012).

In fact, the boosting effect of the carbon dioxide may explain the absence of bacterial growth if this gas is totally lacking in the ambient air (Hakalehto and Hänninen, 2012). Therefore, we could anticipate that the increase of carbon dioxide resulting from climate change could have an accelerating effect on the microbiological degradation processes globally.

The ultrasensitive enrichment and detection device PMEU Scentrion® have been used for the monitoring of dairy products (Figure 6.5). This approach has been tested for the screening of milk spoilage and mastitis bacteria. Moreover, it could be used for the detection of *Mycobacterium avium*, which is a potential risk in causing milk spoilage (Table 6.1).

Table 6.1: Mycobacteria in milk (modified from Eltholth et al. (2009).

Mycobacteria (MAP) as a health risk in milk
– MAP (*Mycobacterium avium* subspecies *paratuberculosis*) is a hazardous microorganism transmitted via milk.
– MAP has also been suggested as a possible cause of Crohn's disease (CD).
– *Mycobacterium* sp. strains are potential clinical pathogens, as well as environmental and food contaminants.
– Pasteurized milk can get contaminated and spread the MAP.
– "Currently available data suggests that the livelihood of dairy and meat products being contaminated with MAP on retail sale should not be ignored."
– Novel methods for monitoring milk hygiene, as any food hygiene issue, need to be incessantly developed.

Early and accurate detection and identification (verification) of harmful bacteria is the key for selecting suitable antibiotics and for the prevention of the spreading of antibiotic-resistant bacterial strains. COVID-19 viral pandemic has raised the awareness of the need for rapid microbial detection in order to respond effectively to different biological threats, such as the antibiotic-resistant bacteria potentially rising up in the aftermath of the viral disease or pandemics. This is particularly important regarding the bacteria causing water and food contamination. Early recognition of multi-resistant or otherwise hazardous bacteria is a prerequisite for logistical hygiene control, prevention, and targeted treatments of the infections. If hygienic precautions fail, these pathogens may spread, for example, via food and drinks, soil, vegetation, animals, air, or contaminated water.

The PMEU has been used also in field conditions in different environments and for various sample matrices (Pitkänen et al., 2009; Hakalehto, 2010; Humppi et al., 2016) enabling also the on-line monitoring of the surroundings (Hakalehto and Humppi, 2018). The PMEU has been used for pre-enrichment of samples originating from foods, environment, and clinical patients (Hakalehto, 2011a; Hell et al., 2015). When needed, it is also possible to screen antibiotic susceptibilities during the pre-enrichment.

The PMEU has also been used in combination with ChemPro® 100i (Environics Oy, Mikkeli, Finland), a sensor for vaporized molecules based on the ion mobility spectrometry method in order to shorten the detection time. When these methods were used in conjunction, bacterial growth was detected in 4–5 h (Hakalehto et al., 2009).

6.7 Hygienic and health effects of food

6.7.1 Internal and external microbes and the balance with the host system

The human digestive microflora consists of (a) the strains coming in within the food, and (b) the microbiome which already exists in the alimentary tract. The latter consists of epithelial biofilms for a large part (Armon, 2015). The local balance of the biofilms is important for the continuity of this community, but it is essential also for the digestive health of the human host.

Molecular communication is continuously existing between the microbes and the mucosal cells. The human body system also secretes enzymes, hormones, acids and basic substances, antibodies, signal molecules, polymers, molecular wastes, bile substances, and other chemical compounds that influence the microbiome.

The microbial cells that reside in our mouth, naso-pharyngeal tract, esophagus, gastric areas, intestines, and entire digestive tract, produce their extracellular secretions. They have effects on other microorganisms but also on the host epithelium. Since the microbiome usually strives for balance (Hakalehto, 2012a, 2020), it naturally extends this demand or objective for balance to the human tissues. Those, in turn, have their homeostasis, which is in molecular communication with the microbiome balances, such as the Bacteriological Intestinal Balance (BIB) (Hakalehto, 2011a, 2012a, 2012b, 2013a). Hence, we as humans are integrated with our microbiome in many ways.

Consequently, the food which we consume is actually processed together with our microbiome, before it is absorbed by our body system. As a result, many of our nutrients are the metabolic products of our microbes. This delicate balance between the human body, its digestion, and the digestive microbes, is the foundation of our individual health.

Any intruding external microbial strain encounters and challenges the built-in stability of the microbiome in various parts of our body system and digestive tract. During the first 5–6 childhood years, our immune system gives "a free ticket" for those bacterial and other strains which are not treated as extruders but are allowed to attach onto the mucosal surfaces and for the delicate and sophisticated community of biofilms on those epithelia (Konto-Ghiorghi et al., 2009; Pesola and Hakalehto, 2011). Viruses do not have any metabolism of their own, and therefore they are not "founding members" of the microbial communities or not even the full members at all, but they do have an important role in the joined infection mechanisms, for instance (Domínguez-Díaz et al., 2019). However, viruses have many other important roles in the biosphere, too. They are specific to their hosts whether human, bacterial, fungal, or other organisms. By infecting cells, they deliver messages also on the population level (Kojima et al., 2021).

As a "sum effect" of the network of interactions, which surround the food uptake and nutrient absorption, it is worth understanding that the integrated microbiome together with the body system strives for maintaining balance, which is a changeable concept depending both on the variations in the food materials and its other components but also on the metabolic potential of the individual microbiome in question.

It is noteworthy that any microbiome strives for balance. And the best way to achieve this balance is the maintenance of the genetic and metabolic versatility of the microbial community. In the studies for describing this structure of microbiome in the light of metabolomic observations, such methods as nucleic magnetic resonance (NMR) could be used (Laatikainen et al., 2016). Therefore, it is well possible that we cannot isolate any specific strains representing a group of microbes unless we first provide the human food or nutrient that enriches this microbe of a certain specificity. In other words, we can use the diet as a selective factor for various microflora. The task of the viruses could also be the transfer of the required or additional genes or functional sequences between related microbial species. This may sometimes be a less common phenomenon *in vivo*, but it could occur in a densely populated community.

In practice, it is highly important to understand the effects of various foods on the microbiome. Consequently, we can deduce their effects on a human individual. Furthermore, if the diet is versatile and the microbiome a diverse one, it is more likely that the human individuals and their immune system can tolerate or even eradicate the harmful or pathogenic strains, or pandemic microbial strains. This ruling out of "troublemakers" could take place in close coordination with the host immune system and other defenses. The gut microbiome balance is affected by several factors, such as additives or dyes in the food. We do not know enough about these interactions, but this does not make them inexistent or inefficient. On the contrary, we can see in the lab and in the clinic, that patient cases speak for themselves. The ameliorated patient condition often relates to the improved balance of the microbiome.

This balance could be achieved in many ways, such as changes in the diet, probiotics and prebiotics, medical treatments, complementary medicine means, fecal microbial transplant, or with the aid of so-called passive immunization. Most importantly, the balance derives from the ecological succession of microbial species as implied by the BIB (Bacteriological Intestinal Balance) (Hakalehto, 2012a). The food can act as a medicine (Hakalehto, 2020), but we may add some balancing molecular activities into it, such as the IgY antibodies (Hakalehto, 2021a). It is important to get enough natural health protection among the food, too. Vitamin D is an example of a molecule, which is highly important for the immune system (Chang and Lee, 2019; Charoenngam and Holick, 2020).

The passive immunization with the egg yolk IgY antibodies has proven out to help patients in dangerous situations, threatened by hazardous infections, dysbiosis, or other serious microbiological imbalances. An example of the process for producing IgY is presented in Figure 6.7. This schematic presentation was published for the first time in the Finnish Kemia magazine on June 18, 2020. Later, it was translated into English and integrated into the application to the European Union in Summer 2020 in order to establish experimental or modular production of the IgY antibodies. This application was submitted as a joint application by many classy research institutes and universities, such as Sheba Medical Center of Tel Aviv, Israel, which was ranked as the ninth best hospital in the world by Newsweek magazine. The Wroclaw Medical University of Poland, the Universities of Uppsala, Stockholm,

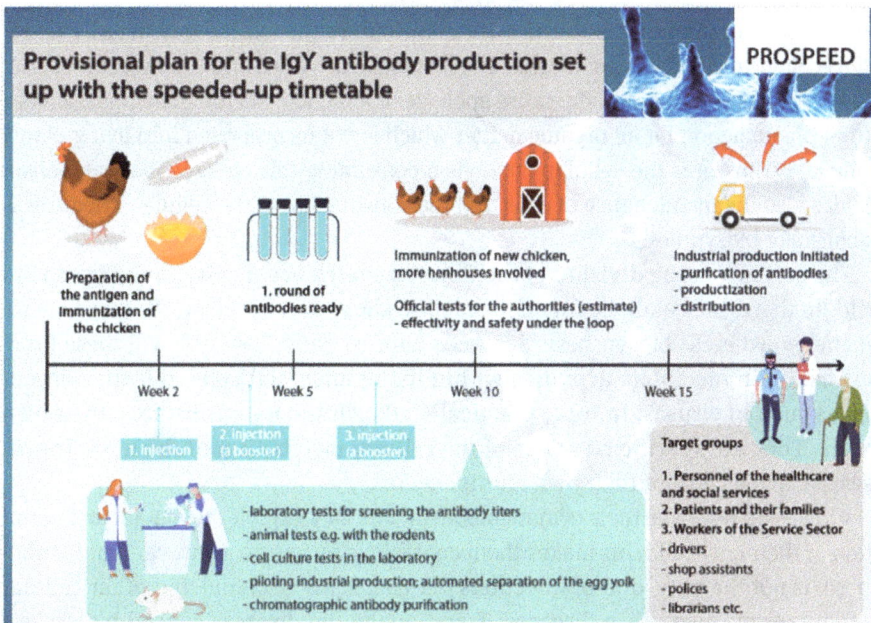

Graphic: Viivi Räsänen © Elias Hakalehto & Finnish Chemical Magazine

Figure 6.7: Schematic plan for the prophylactic IgY antibody production in chicken eggs (Kemia Magazine, June 2020).

and Mälardalen of Sweden, as well as the Virology Department of the Tampere University in Finland and the Finnish Center for Vaccinations in Tampere. Regardless of the competent partners, European Union decided not to admit a grant (Kemia-Magazine, August 2020).

Besides the academic institutions and hospitals, also several private companies were involved in the above-mentioned application process to the European Union. The lead partner was Finnoflag Oy and the process was conducted by Adj. Prof., microbiologist Elias Hakalehto. Several international studies had evidenced the safety and efficiency of the IgY method. For example, the Uppsala University Hospital in Sweden had used IgY and passive immunization for at least 20 years to protect some cystic fibrosis patients against *Pseudomonas aeruginosa* infections in the lungs (Larsson and Carlander, 2003). The application co-operation in Sweden was coordinated by Senior Professor Erik Dahlquist of MDH, Västerås. Other positive evidence has been obtained all over the world (Nguyen et al., 2010; Pérez de la Lastra et al., 2020). This kind of use of food components could be associated with the use of food as medicine (Hakalehto, 2020).

6.8 Further roles of viruses and microorganisms

The functions of microbes in Nature are manifold. In principle, they usually relate to the ecological circulation of matter. The metabolic activities and propagation of bacteria and other microbes actually cause spoilage of food, too. Microbial degradation is a natural phenomenon for all organic matter which is not incorporated into living plants or animals. Thus, it is the cellular catabolism consuming this energy. That catabolism includes also the maintenance of cells. But the construction of the cellular structures is anabolism or biosynthesis.

The corresponding division of the flows of energy occurs also in the microbial world. In there, the catabolic functions require about 85% of the metabolized chemical energy, whereas biosynthesis demands approximately 15% of it. All these functions maintain metabolic activities within living microbial cells, but they do not exist as such in viruses. In turn, practically all cellular organisms have their own viruses. They are specific parasites of the cellular metabolism of the host. The viruses use living cells for their own multiplication.

Viruses do not have their own metabolism, but they exploit the metabolic organization of their host cell. This makes them contagious by nature. However, viral pathogenesis is not the only kind of infectious disease. Many bacterial and other cellular microorganisms also cause diseases. Traditionally the diseases caused by viruses, bacteria, or other microorganisms can be classified as contagious illnesses caused by specific germs. However, many diseases are caused by simultaneous activities of several distinct organisms, or by a malicious balance of the microbial community.

The prevention of pathogens by vaccination or by using systemic antibodies has evoked apprehension related to the safety issues. In any case, these worries are controversial and have not been demonstrated in the prophylaxis for SARS-CoV-2. However, antibody-dependent enhancement (ADE) of disease is a general concern for the development of vaccines and antibody therapies (Arvin et al., 2020). The mechanisms that underlie antibody protection against any virus have a theoretical potential to amplify the infection or trigger harmful immunopathology.

As scientific research has progressed throughout many decades or for several centuries, ideas of microbial diseases have broadened considerably. For example, many illnesses or syndromes have been discovered to be caused by multiple strains. Also, many human conditions are interrelated with microbial activities which may turn into malicious ones in time or in changing conditions. Thus, the diagnostics of infections has become more complicated, and the consequences of ill health are often called "syndromes." For example, irritable bowel disease (IBD) is often not just one sickness, but merely a combination of several infectious bowel diseases or inflammations.

The viruses are "partially living organisms"; they propagate themselves and produce offsprings of their kind. However, they do not possess cellular metabolism of their own. Therefore, they could be considered as "cast-outs" or "spin-offs" of the genetic organization of their host. And such hosts could be in all the various living cells, including human, animal, plant, and microbial cells. In this role of transducers between cells, the viruses transform genetic material from cell to cell. Their role is undoubtedly an important one in the epigenetic regulation of the genomes (Jaenisch and Bird, 2003). Viruses recognize and cause the formation of novel variants. Thus, it is not meaningless, what type of viruses does our food contain. In fact, they should also be monitored like bacteria and other microbes. It is important to critically investigate the mechanisms of contagion. For example, according to the modern understanding and research, it seems likely that *Yersinia pestis*, a plague-causing agent, was not transmitted by rodents, but by lice, also in the human-to-human transmission (Barbieri et al., 2021). According to novel understanding, it is also likely that some viruses could have paved the way for the bacterial epidemics occurring in their aftermath (Hament et al., 1999).

6.9 Build-up of intestinal microbiome and the entering of foreign strains from food and environment into the joint ecosystem with host system

In our intestines during the early days of infancy, at the moment of birth and after the delivery, numerous microbial strains inoculate and inhabit our skin, intestines, and other epithelia. Our body system hardly produces antibodies by that time of early childhood, against intruders. Instead, we are dependent on the flow of maternal antibodies through the placenta (Niewiesk, 2014) or in the breast milk (Van de Perre, 2003). A proof of the high importance of this supplementation of protectives is the high concentration of IgY antibodies produced by chicken into their eggs for the immunoprotection of the hatching young ones (Hakalehto, 2021a). Since the birds do not have placentas or wombs, they have to produce antibodies into their egg yolk. From there, these molecules move into the epithelia of the newborn, hatched individuals.

In the case of human toddlers, they encounter, besides the bacterial and other microbial strains of their mothers' microbiome, also other microbes of their surroundings. These microorganisms form the normal flora of the human individual, which in a way is "under the radar" of the immune system in the early years of a human individual. Thus, we receive the components of our normal flora during the 5–6 first years of our lives. This development takes place together with the maturation of the immune system (Ganeshan and Chawla, 2014). Hence, it is of high importance that during these years, we can build up a microbiome which is versatile and functional enough, and has established molecular communication with the host immunological, humoral, neuronal, and metabolic networks. In the construction or revival of the intestinal microflora and functional microbiome, it is important to take into account the enterohepatic circulation (Hauser et al., 2006).

In order to avoid dysbiosis, diseases, or inflammatory conditions in the gut, we need to have an active and functional BIB (Hakalehto, 2011a, 2012a, 2012b, 2013a; Hakalehto et al., 2008, 2010). This delicate balance has its foundations in the uppermost parts of the small intestines, in the duodenum. The various groups of "coliforms" belonging to the family *Enterobacteriaceae* contribute to that balance (Hakalehto et al., 2008, 2010; Pesola and Hakalehto, 2011). The mixed-acid producing strains of *E. coli* and its relatives are the fastest reacting strains as the food enter the duodenum from the acidic stomach. In the duodenum, the pancreatic bicarbonates neutralize the incoming food, making the pH to rise from 1–2 up to 6. As the mixed-acid fermenting strains produce acetic acid, this provokes the *Klebsiella/Enterobacter* group of coliform bacteria to produce more neutral substances, such as ethanol and 2,3-butanediol, which also take part in the neutralization process and metabolic balancing of the digestive tract (Hakalehto et al., 2010).

In the case of newborn children, the BIB seems to develop naturally on the basis of the balance between the above-mentioned two groups of coliform bacteria, namely the mixed-acid fermenting and 2,3-butanediol producing ones (Hakalehto et al., 2008; Pesola et al., 2009; Pesola and Hakalehto, 2011; Hakalehto, 2012a,b). Concomitantly, they also form mixed cultures, which inhabit the first sections of the intestinal region and form a balanced microbiological foundation of the microbiome (Hakalehto, 2012a, 2012b, 2018). These mixed populations are then kept stable by the human molecular mechanisms, such as antimicrobial peptides (Hakalehto, 2011a) or by the lactic acid bacteria (LAB), which traverse through the entire course of the intestines along with the chyme (Hakalehto, 2011a, 2015a). The LAB also attenuates the excessive growth of the coliforms (Hakalehto and Jaakkola, 2013). From the human point of view, it should be remembered that several microbes have anticarcinogenic effects on our body (Havenaar, 2011; Nakkarach et al., 2021).

As the food transforms into chyme and is passing through the small bowel, 80% of the total nutrients for human consumption are taken up. They are absorbed during that period of 6–7 h. The circulation of bile substances takes place simultaneously, and they have an impact on the microbial constitution (Hakalehto et al., 2010). However, when the chyme enters the cecum, its microbiological composition of species and strains is changing radically, and the numbers of microbial cells are increasing in numbers. In any case, these microbes continue their interaction, as well as the molecular communication with the host epithelium. For example, the carbon dioxide emission of the lactobacilli and others is boosting the butyric acid production by clostridia (Hakalehto and Hänninen, 2012). The butyric acid, in turn, is important for the health of the human large intestines and for the regulation of water balances in the body (Velázquez et al., 1997). In summary, the intestinal microbiome has crucial tasks and an important contribution to our health in many ways (Figure 6.8).

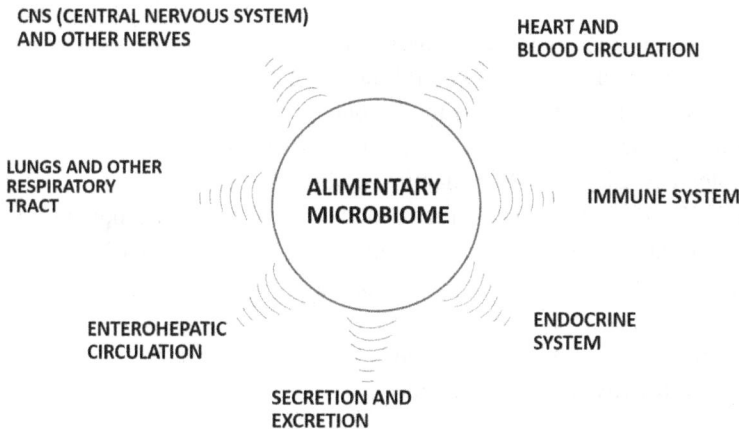

Figure 6.8: Effects of the alimentary microbiome and its functions are reaching to all parts of the human system and its metabolism (Hakalehto, 2021b).

6.10 Antibiotic-resistant bacteria

The spreading of new antibiotic-resistant bacteria is a substantial problem both in welfare societies and in all geographical areas. Developing new antibiotics is inefficient and expensive. Because of that, there are only limited resources focused on this R&D activity, which should be increased for essentially saving human lives.

In animal husbandry, antibiotics are used for preventing diseases and sometimes also for making the beef cattle or other animals grow more rapidly. However, the use of antibiotics as growth-promoting agents is nowadays forbidden in many countries (Low et al., 2021). With modern technology, antibiotic contamination from animal husbandry can be tracked anyhow (Economou and Gousia, 2015). Antibiotics have been a functional solution for curing many clinical bacterial infections for almost a century. However, the excessive use of antibiotics (not only in the clinical field but also in animal husbandry and agriculture) has led to the emerging of new resistant strains. This undesired trend has been influenced also by the dissemination of resistant bacteria in the hospital and municipal wastewaters (Wang et al., 2019). Still, the wise use of antibiotics could be an answer for many problems involving the declined sensitivities of bacteria toward antimicrobial medication.

There are several methods for reducing antibiotic resistance such as molecular biology techniques, finding the minimal inhibitory concentrations or precise specifications of the antibiotics, or inhibitory zones obtained with the disc diffusion method (Figure 6.9). Restrictions in the use of antibiotics in animal production could be the key to decrease the spreading of antibiotic resistance, too. In order to achieve fast and reliable information on the bacterial susceptibilities of various antibiotics, such methods as enhanced enrichment with the PMEU could be used (Hakalehto, 2011b).

In the United States, a study found an association between heavy metal tolerance and antimicrobial resistance. This could mean that the metal cations might prevent antibiotic substances from entering the bacterial cells (Chen et al., 2019). This indicates the relationships between various parameters. It also implies the increased microbiological risks caused by chemical pollution.

In 2011, the US FDA's Center for Veterinary Medicine published a report stating that 50% of all retail meats contained markers of antibiotic resistance (Gebeyehu, 2021). This could mean that groceries could be among the most significant sources for the dissemination of antibiotic resistance. In a related study, it was found out that the usage of antibiotics in milk production is the principal reason for antibiotic resistance in dairy industries (Sachi et al., 2019).

The extensive use of antibiotics in agriculture has proven out to be the major cause for the selection and spreading of antibiotic resistance (Suriyasathaporn et al., 2012). There is a variety of ways for resistant strains to be distributed throughout the environment. Bacteria can spread antibiotic resistance genes to other bacteria by transfection, plasmid conjugation, or transformation. It is also possible that the antibiotic resistance markers spread from some bacterial cells to the adjacent strains through the common metabolic actions and interactions of the microflora.

Antibiotic resistance is often associated with bacteria and other microbes living in the natural soil environment (Cycoń et al., 2019). In the soil ecosystem, the restricted nutrient availability and other conditions create competition for survival between different microbial strains. In this ecosystem, many bacteria exploit their capabilities to

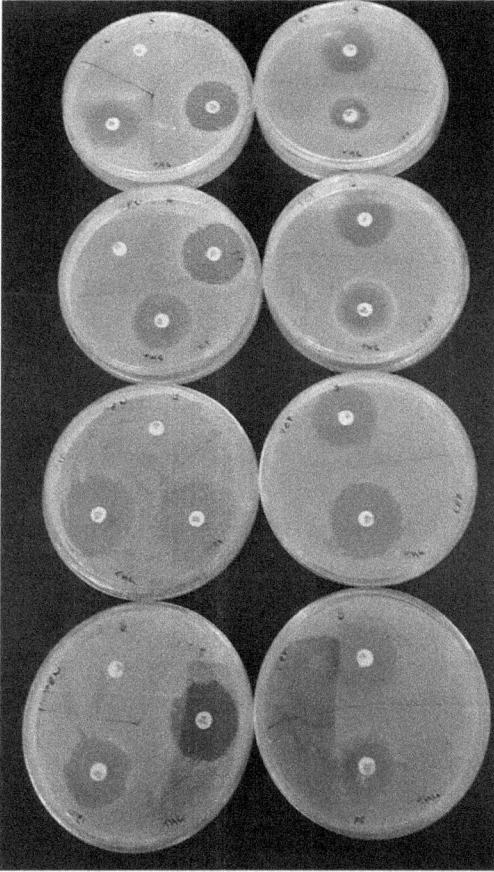

Figure 6.9: Antibiotic resistance monitoring on the Petri dishes (TYG agar) with inhibition zones around the antibiotic discs. The bacterial strains from up to down are patient strains of an individual infant: 5 *Enterobacter cloacae* (IIc4), 6 *Enterobacter cloacae* (IIc5), 7 *Escherichia coli* (Ia1), 8 *Escherichia coli* (Ia2). Antibiotic discs on the bacterial carpet, as follows (codes in parentheses): penicillin (PEN), piperazine (PIP), trimethoprim (TRI), tetracycline (TET), cephalosporin (CEF). *Enterobacter cloacae* is an enteric species often found in clinical blood transfusion equipment, catheters, and so on. It also shows a tendency, partially due to this niche in the hospitals as well as its growth habits and biofilm formation, to form antibiotic-resistant strains (Hakalehto, 2011b, 2015f). The extensive emergence of bacterial antibiotic resistance is one of the major threats associated with the long-term consequences of COVID-19 pandemics. Photograph taken by Anneli Heitto, Finnoflag Oy laboratory.

produce antibiotics to maintain their niches. However, the environmental circumstances for bacteria in soil are different from the ones in the human digestive system. If the microbial strains are exposed together to a specific environmental niche for a sufficient time, the optimal conditions occur for the growth of this mixed population.

In healthcare, new problems have emerged with respect to bacterial infections and their treatments. Many bacterial strains are resistant to several if not most antibiotics. These strains are called multi-resistant ones. Their occurrence implies uncontrolled or reckless usage of antibiotics. It drives us to consider old primary methods in contracepting bacterial infections. Already in the late 1800s in Vienna, Hungarian Ignac Semmelweiss found out that a lime solution that contained chlorine was able to remove the displeasing smell from the hands of doctors after an autopsy. The cleaning of hands of the care personnel also turned out to lower the risk of puerperal fever. This principle of handwashing has not lost its effectiveness in the battle against bacterial infections. It is crucial because most of the patients in healthcare facilities have lowered resistance toward pathogens. The so-called Semmelweiss' principles are still valid in today's nursing and medical care (Hakalehto, 2006; Laitiomäki et al., 2015; Hakalehto et al., 2021).

Industries, agriculture, and healthcare units are all transmitting antibiotic-resistant bacteria into the environment. Many studies have indicated that in healthcare facilities, the extensive use of antibiotics and carelessness in the basic principles of infection control have aggravated antibiotic resistance (Gil-Gil et al., 2021). In this alarming situation, to prevent contamination and dissemination of harmful or hazardous bacteria, the whole clinical environment should be monitored, including its equipment and personnel. This procedure should be performed following proper sampling protocol and convenient enrichment techniques.

One reason for the high tolerance that many bacteria have shown in extreme conditions is the spores they are able to produce. This capability is associated with the aerobic genus *Bacillus* and the anaerobic *Clostridium* in particular. Some bacterial spores have been shown to resist drying and even boiling for relatively long periods of time. The molecular communication between bacterial cells makes it possible for them to form different structures to survive difficult environmental stresses.

In the search for an effective solution for the prevention of the selection of antibiotic-resistant bacteria, studies have been performed regarding the old, traditional methods in preventing contamination. For example, it has been found out that frogs secrete antibiotic peptides from their skin (Samgina et al., 2012, 2016). This has been used to keep the water in drinking wells clean and safe, thanks to the magainin peptides of the amphibians. Also, many other sources for the novel antimicrobial substances have been proposed (Hayashi et al., 2013; Silva et al., 2011).

In the 1990s, an experimental study about antibiotic sensitivities of a food contaminating *Pectinatus* sp. was made in the Department of Membrane and Ultrastructure Research at Hadassah Medical School of the Hebrew University of Jerusalem, Ein Kerem, Israel. The study was conducted by Elias Hakalehto and the late Prof. Itzhak Kahane, the former chairman of the Israeli Society of Microbiology (Hakalehto, 2015f). The study evidenced that in many breweries the same strain of *Pectinatus* sp. contaminant can survive in unchanged conditions for years. Alternatively, many strains were present with reminiscent capacities. This research raised ideas on how the said anaerobic bacteria could survive in factory conditions. This study also demonstrated the fact that food and natural substances could be useful in finding new sources of antimicrobials. The similarities in hygienic maintenance operations in the food production and healthcare field connect these fields with each other for the search and elaboration of novel pharmaceuticals.

In the following, the current issues of antibiotic resistance are discussed in the blogs of Hakalehto E. Probiotics, Antibiotics and Politics, published in UVC-LED BLOG & NEWS. LED FUTURE, on April 6, 2021 (Hakalehto, 2021d):

> Growing numbers of antibiotic-resistant bacterial strains have caused great concern for many years. Now they really should be confined better in the aftermath of the pandemics, which has downgraded the immunological strength and epithelial defenses of many, thus enabling the emerging bacterial epidemics to pop up. – Extended-spectrum beta-lactamase (ESBL)-producing *E. coli* was found to be present in 11% of human fecal samples in all Britain in 2013–2014 and 17% of the corresponding specimens in London, according to a report published in Lancet Infectious Diseases online in 2019 by National Infection Services, other British authorities and some universities in England, Scotland and Wales. In another survey just before the SARS-CoV-2 pandemics, carbapenemase-producing *Enterobacteriaceae* (CPE) in England between May 2015 and March 2019, as many as 30.3% of the *E. coli* isolates and 39.1% of the *Klebsiella* sp. isolates, were detected as CPE positive. This work was published in "Infection Prevention in Practise" n:o 2 (Mawdsley, 2020). https://journals.sagepub.com/doi/10.1177/1757177420935633

> In India, 13.2% of the population received the first dose of the anti-SARS-CoV2 vaccine by 6th of June 2021. However, as an alarming sign of the potential complications, Mucormycosis (the "black fungus") has spread among the diseased. This dangerous fungal infection of the respiratory tract has become increasingly abundant due to high SARS-CoV-2 infection rates. Such problems have sometimes occurred also in the water-damaged buildings in Finland and elsewhere. This demonstrates the need to shelter from the conjugated pathogenesis, not against a single causative agent only. At the end of the episode, it might happen that the risk taught to be a secondary one becomes the worst long-term consequence. (Hakalehto E. Pandemics, mycobacteria and climate change – risks for emerging epidemics in the light of accelerating climate change. Blog published in Microbiome Power (ed. Jakovljevic V.) on June 8, 2021. https://microbiomepower.com/2021/06/08/risk-for-emerging-epidemics-mycobacteria-and-accelerating-climate-change/)

6.11 PMEU – equipment for food monitoring

PMEU (Finnoflag Oy, Kuopio, Finland; Hakalehto et al., 2009, 2011) is a portable incubator where the enrichment of microbes takes place in syringes containing enrichment broth (Figure 6.4). See also Chapter 4 of this book. Gas flow is directed into syringes through a sterile filter and a needle to agitate the broth and the bacterial cells. Different bubbling gases can be used in aerobic and anaerobic enrichment, respectively (Hakalehto et al., 2013a,b).

The PMEU has been designed for promoting the recovery and to accelerate the growth of microbes under an aseptic gas flow in various adjusted or programmed temperatures (Hakalehto et al., 2009). The PMEU units are remote-controlled. They reveal the enteric and other bacteria in clinical, industrial, and environmental samples and have also been used in the investigations regarding the properties of enteric bacteria and their interactions between the other members of the intestinal flora (Hakalehto, 2006, 2011a; Hakalehto et al., 2007, 2008, 2009, 2011, 2013a; Pesola et al., 2009; Pitkänen et al., 2009). The PMEU method has been validated for the detection of coliform bacteria (Wirtanen and Salo, 2010) and for the enrichment and detection of *Campylobacter* sp. in waters (Pitkänen et al., 2009).

In the food hygienic monitoring applications, PMEU Scentrion® version was used for screening the artificial contamination of chicken meat (Figure 6.10). See also Hakalehto et al. (2015a).

naturally spoiled chicken meat (kept at room temperature)

artificially contaminated chicken (with a strain of Salmonella enterica Serovar Typhimurium)

Figure 6.10: Food hygienic *Salmonella* monitoring of chicken meat. The peak of gas emission as measured by the MOS2 sensor of the Chempro 100i™ unit (Environics Oy, Mikkeli, Finland) of the PMEU Scentrion® device (Finnoflag Oy, Kuopio, Finland) for the rapid detection of microbial contamination. Modified from Hakalehto et al. (2015a).

6.12 Botulism – essential risk hiding in many foods

Clostridium botulinum is a sizeable Gram-positive, anaerobic, sporulating bacillus-type of bacterium that can be found in soil, freshwater, saltwater, wastewaters, and insects (Pesola et al., 2015).

C. botulinum can synthesize botulinum neurotoxins (BoNT) that belong to the most toxic compounds found in nature. Seven different BoNT types, namely types A–G, vary regarding their antigenic structures and characteristics. The lethal dose (LD) of BoNT for humans is approximately 10 pg/kg of body weight.

The toxic effect of BoNT is achieved in three phases (Bandyopadhyay et al., 1987; Maisey et al., 1988; Sathyamoorthy et al., 1988; Pellizzari et al., 1998):

1. C-terminal component of the H chain (H1) binds to a receptor located in the nerve ending.
2. N-terminal component of the H chain (H2) opens a tunnel to the cell membrane of the target cell for the toxin to enter the cell.
3. The L chain is a zinc-protease that inhibits the release of neurotransmitters from the cell by specifically cleaving three proteins participating in the exocytosis process of the transmitters.

None of the toxin components is toxic when alone, and all the components have to be available in the above-mentioned numerical order to gain their function as poison (Maisey et al., 1988).

The most common source of food-borne botulism is fish, but vacuum-packed and canned foods and honey are also often contaminated. The *C. botulinum* bacterium and its toxins have to be always taken into account due to their severity as pathogenic agents, as well as for their high stability in foods.

The BoNT's enter the human body usually through the mucous membranes of the airways or via the gastrointestinal tract. Afterwards, the symptoms appear in 2–72 h. The BoNTs bind irreversibly to cholinergic synapses of peripheral nerves. This is exhibited, for example, in myoneural junctions, and the binding prevents the release of the acetylcholine transmitter (Binz et al., 1990). This leads to flaccid paralysis, propagating muscle by muscle, and causes ultimately the death of the patient as a consequence of the paralysis of respiratory muscles and myocardium. Mortality rate of severe BoNT poisoning is as high as 60–70%.

The *C. botulinum* spores can resist even hours of boiling, but their inactivation can be performed by pressure sterilization. *C. botulinum* strains can produce the BoNTs in as low as +3.3 °C (Jahkola et al., 1999), i.e. even in foods stored in the refrigerator. The BoNTs get degraded by boiling at +100 °C for 10 min or in the open air for approximately 12 h. Based on this information the cold chain of vacuum-packaged foodstuff is essential in the prevention of botulism. On the other hand, these foodstuffs should be cooked long enough.

The production of BoNTs appears to be a defensive reaction of the bacterial cells when they are exposed to disadvantageous environment. On the other hand, the production of BoNTs can be reduced by nitrogen-containing molecules, like arginine, glutamine, or tryptophane amino acids and ammonia, that induce the well-being of *C. botulinum* bacteria (Leyer and Johnson, 1990).

When BoNT food poisoning is suspected, the absorption of the toxin should be minimized, for example, by inducing vomiting and by performing intestinal lavage. The treatment of botulism is based on the support of vital functions. Respiration and blood circulation are supported when needed. Penicillin antibiotic treatment is used in wounds contaminated by *C. botulinum*. However, in other types of botulism, antibiotic treatment is not advised, because it can speed up the release of toxins (Mayo Clinic, 2020). Although a bactericidal type of antibiotic substance can destroy the bacteria, it cannot deactivate the toxins already released. BoNTs can be neutralized by specific antitoxins, for example, commercially available antitoxin-ABE, that is produced in the horse. Antitoxin-ABE refers in here to Trivalent Botulinum Antitoxin Type A+B+E. Also, botulism immune globulin sera can be used for the treatment of infants.

For immunological detection of BoNTs, we synthesized a peptide of 14 amino acids according to the amino acid sequences of BoNT proteins (Pesola et al., 2015). This peptide was used for the immunization of rabbits to induce production of antibodies against the toxins. The immune reactions were measured by enzyme-linked immunosorbent assay (ELISA) (Absorbance of 405 nm, 1 h). The mean value of the titers of rabbit immune sera able to detect the BoNT peptide was 1:4,500 after 4 immunizations and 1:8,000 after 7 immunizations.

The rabbit immune sera were also used for the detection of BoNT peptide that was mixed with different foods: minced beans (pH about 6), soured pasteurized cow's milk (pH 7.19); and environmental samples: distilled water (pH 4.5) and lake water (pH 8.14). The immunological detection of BoNT peptide was successful from distilled water and lake water, whereas from beans and cow milk it was remarkably hindered, probably due to various bioactive molecules of these foods (Pesola et al., 2015). Hence, there is a need for further development of the method to enable the detection of BoNT peptides also from samples with high protein contents or otherwise complicated matrices.

6.13 Conclusion

The microbiological risks in food comprise food spoilage, the contagion of diseases, toxin formation, impact on the gut microflora due to food microbes, and so on. Anyone of these risks is very serious and life-threatening as such. During pandemic times, they are readily associated with the risks of getting infected by the viruses or bacteria spreading around.

The procedure for developing antimicrobial, antiviral, or antitoxin antibodies based on peptide sequences, is a safe and successful approach for the diagnostic purposes. The eradication or prevention of toxin formation by spoilage bacteria could serve in the avoidance of any food-related risks. However, the toxins are not propagated in food without the bacteria producing them. Correspondingly, viruses always need their host cells for multiplication. The methods for the detection of toxins and other antigens from the sample matrices with high protein content need to be developed. Also the genetic methods, such as PCR, are prone to both false negative and false positive results. However, they constitute the fastest procedures to detect, identify and characterize specific nucleic acid sequences.

Research on various antibodies and autoimmune reactions, as well as genetic tests, in the background of several symptoms of COVID-19 and other microbial diseases should be intensified. Whatever measures are taken into use for the prevention of pandemic events, or any other microbiological or toxicological hazards, they have to be scientifically justified and take place with correct timing and in accordance with the needs of various patients of different population groups. The education of the general public has proven important during the current COVID-19 pandemics. Additionally, such specialists as food or healthcare workers need more microbiological education. The rapid development of vaccines during the current pandemics has provoked worries about unexpected or undesired risks, such as the ADE (Antibody Dependent Enhancement).

The constant increase in the number of antibiotic-resistant bacteria should clearly point out the necessity and urgency to protect the coronavirus-strained population against the risks of these resistant bacterial variants potentially occuring in the aftermath of the pandemics. During the "Spanish flu" a century ago, about 70% of the deaths connected with the virus pandemics were associated with bacterial infections. We cannot afford poor health politics now to allow the spread of these devastating modern-day threats. Also, the food production and distribution chains need to take still more responsibility. The developmental efforts to find novel antibiotics and probiotic strains have to be accelerated and intensified, and increasingly resourced. Also, the improved hygiene levels during hospital care, surgery, out-patient treatments and antibiotic therapies have to be emphasized. New medical devices and protective means such as UVC (Ultraviolet-C) light equipment need to get implemented. Environmental pollution has to be eliminated as a source of infections or allergies by novel, innovative techniques.

Finally, we need to understand the microbiome as our internal ecosystem, the BIB in it, as well as our links by the microflora with the environment for better health and happiness. The strategies of using food hygiene control and quality improvement (of cleanliness and nutritive values) have to lead to the betterment of individual health. We need to position ourselves with novel strategies also for the fight against new emerging pandemics. This includes improved understanding, regarding the exchange or spread of microbes between human individuals, animals, and the environment. The microbiological aspect or approach of life could lead us to happier and healthier societies, where novel innovations improve the standard of living of the citizens.

References

Alder-Rangel, A., Bailão, A. M., Da Cunha, A. S., Soares, C. M. A., Wang, C., Dadachova, E., Hakalehto, E., Eleutherio, E. C. A., Fernandes, E. K. K., Gadd, G. M., Braus, G. H., Braga, G. U. I., Goldman, G. H., Malavazi, I., Hallsworth, J. E., Takemoto, J. Y., Fuller, K. K., Selbmann, L., Corrochano, L. M., Von Zeska Kress, M. R., Bertolini, M. C., Schmoll, M., Pedrini, N., Loera, O., Finlay, R. D., Peralta, R. M., Rangel, D. F. N. (2018). The second International Symposium on Fungal Stress: ISFUS. Fungal Biology. Jun;122 (6): 386–399. doi: 10.1016/j.funbio.2017.10.011. Epub 2017 Nov 20.

Armon, R. (2015). Biofilm formation in food. In: Hakalehto, E. (ed.). Microbiological Food Hygiene. New York, NY, USA: Nova Science Publishers, Inc.

Arvin, A. M., Fink, K., Schmid, M. A., Cathcart, A., Spreafico, R., Havenar-Daughton, C., Lanzavecchia, A., Corti, D., Virgin, H. W. (2020). A perspective on potential antibody-dependent enhancement of SARS-CoV-2. Nature. 584 (7821): 353–363. doi: 10.1038/s41586-020-2538-8. Epub 2020 Jul 13. PMID: 32659783.

Bandyopadhyay, S., Clark, A. W., DasGupta, B. R., Sathyamoorthy, V. (1987). Role of the heavy and light chains of botulinum neurotoxin in neuromuscular paralysis. The Journal of Biological Chemistry, 262: 2660–2663.

Barbieri, R., Drancourt, M., Raoult, D. (2021). The role of louse-transmitted diseases in historical plague pandemics. Lancet Infectious Diseases. 21 (2): e17–e25. doi: 10.1016/S1473-3099(20) 30487-4. Epub 2020 Oct 6. PMID: 33035476.

Binz, T., Kurazono, H., Wille, M., Frevert, J., Wernars, K., Niemann, H. (1990). The complete sequence of botulinum neurotoxin type A and comparison with other clostridial neurotoxins. The Journal of Biological Chemistry, 265: 9153–9158.

Carr, M. A., Thompson, L. D., Miller, M. F., Ramsey, C. B., Kaster, C. S. (1998). Chilling and trimming effects on the microbial populations of pork carcasses. Journal of Food Protection, 61: 487–489. doi: 10.4315/0362-028x-61.4.487.

Chang, S. W., Lee, H. C. (2019). Vitamin D and health – The missing vitamin in humans. Pediatrics and Neonatology, 60 (3): 237–244.

Charoenngam, N., Holick, M. F. (2020). Immunologic effects of vitamin D on human health and disease. Nutrients, 12 (7): 2097. Published Jul 15. doi: 10.3390/nu12072097.

Chen, J., Li, J., Zhang, H., Shi, W., Liu, Y. (2019). Bacterial heavy-metal and antibiotic resistance genes in a copper tailing dam Aaea in Northern China. Frontiers in Microbiology, 20 August 2019. doi: org/10.3389/fmicb.2019.01916.

Cycoń, M., Mrozik, A., Piotrowska-Seget, Z. (2019). Antibiotics in the soil environment – degradation and their impact on microbial activity and diversity. Review article. Frontiers in Microbiology, 08 March 2019 |.doi: https://doi.org/10.3389/fmicb.2019.00338.

Dasgupta, S., Crunkhorn, R. (2020). A history of pandemics through the ages and the human cost. The Physician 6 (2) pre-print v1 ePub 29.05.2020. DOI: 10.38192/1.6.2.1

de Chadarevian, S., Raffaetà, R. (2021). COVID-19: Rethinking the nature of viruses. History and Philosophy of the Life Sciences. 43, 2. https://doi.org/10.1007/s40656-020-00361-8

de Wit, E., van Doremalen, N., Falzarano, D., Munster, V. J. (2016). SARS and MERS: Recent insights into emerging coronaviruses. Nature Reviews Microbiology, 14 (8): 523–534. doi: 10.1038/nrmicro.2016.81. Epub 2016 Jun 27. PMID: 27344959; PMCID: PMC7097822.

Domínguez-Díaz, C., García-Orozco, A., Riera-Leal, A., Padilla-Arellano, J. R., Fafutis-Morris, M. (2019). Microbiota and its role on viral evasion: Is it with us or against us? Mini Review. Frontiers in Cellular and Infection Microbiology. Published on 18 July 2019. doi: 10.3389/fcimb.2019.00256.

Economou, V., Gousia, P. (2015). Agriculture and food animals as a source of antimicrobial-resistant bacteria. Infection and Drug Resistance, 8: 49–61. doi: https://doi.org/10.2147/IDR.S55778

Eltholth, M. M., Marsh, V. R., Van Winden, S., Guitan, F. J. (2009). Contamination of food products with *Mycobacterium avium paratuberculosis*: A systematic review. Journal of Applied Microbiology. 107 (4): 1061–1071. doi: 10.1111/j.1365-2672.2009.04286.x. Epub 2009 Mar 30. PMID: 19486426. ISSN 1364-5072.

Freund, J., Casals, J., Page Hosmer, E. (1937). Sensitization and antibody formation after infection of tubercle bacilli and paraffin oil. Proceedings of the Society for Experimental Biology and

Medicine. Society for Experimental Biology and Medicine (New York, N.Y.), 37 (3): 509–513. doi: org/10.3181/00379727-37-9625.

Friedhoff, R. A., Houben, A. P., Leblanc, J. M., Beelen, J. M., Jansen, J. T., Mossel, D. A. (2005). Elaboration of microbiological guidelines as an element of codes of hygienic practices for small and/or less developed businesses to verify compliance with hazard analysis critical control point. Journal of Food Protection. 2005 Jan;68 (1): 139–145. doi: 10.4315/0362-028x68.1.139. PMID: 15690815.

Ganeshan, K., Chawla, A. (2014). Metabolic regulation of immune responses. Annual Review Of Immunology, 32: 609–634. doi: 10.1146/annurev-immunol-032713-120236. PMID: 24655299; PMCID: PMC5800786.

Gebeyehu, D. T. (2021). Antibiotic resistance development in animal production: A cross-sectional study. Veterinary Medicine (Auckland), 12: 101–108. doi: doi.org/10.2147/VMRR.S310169.

Geggel, L. (2020). The world's biggest dust bunny is crossing the Atlantic Ocean right now. Livescience. June 24, 2020.

Gil-Gil, T., Ochoa-Sánchez, L. E., Baquero, F., Martínez, J. L. (2021). Antibiotic resistance: Time of synthesis in a post-genomic age. Computational and Structural Biotechnology Journal, 19: 3110–3124. Published 2021 May 21. doi: 10.1016/j.csbj.2021.05.034.

Haikara, A., Penttilä, L., Enari, T. M., Lounatmaa, K. (1981). Microbiological, biochemical, and electron microscopic characterization of a *Pectinatus* strain. Applied and environmental microbiology, 41 (2), 511–517. https://doi.org/10.1128/aem.41.2.511-517.1981

Hakalehto, E. (2000). Characterization of *Pectinatus cerevisiiphilus* and *P. frisingiensis* surface components. Use of synthetic peptides in the detection of some Gram-negative bacteria, PhD Thesis, Kuopio University Publications C, Natural and Environmental Sciences 112, Kuopio, Finland.

Hakalehto, E. (2006). Semmelweis' present day follow-up: Updating bacterial sampling and enrichment in clinical hygiene. Pathophysiology, 13, 257–267. http://dx.doi.org/10.1016/j.pathophys.2006.08.004

Hakalehto, E. (2010). Hygiene monitoring with the Portable Microbe Enrichment Unit (PMEU). 41st R3 -Nordic Symposium. Cleanroom technology, contamination control and cleaning. VTT Publications 266. Espoo, Finland: VTT (State Research Centre of Finland). pp. 164–176.

Hakalehto, E. (2011a). Simulation of enhanced growth and metabolism of intestinal *Escherichia coli* in the Portable Microbe Enrichment Unit (PMEU). In: Rogers, M. C., Peterson, N. D. (eds.). E. coli infections: Causes, treatment and prevention. New York, USA: Nova Science Publishers, 159–175.

Hakalehto, E. (2011b). Antibiotic resistance traits of facultative *Enterobacter cloacae* strain studied with the PMEU (Portable Microbe Enrichment Unit). In: Antonio, M.-V., editor. Science against Microbial Pathogens: Communicating Current Research and Technological Advances. Badajoz, Spain: Formatex Research Center, Microbiology Series N: o3, Vol. 2, 786–796.

Hakalehto, E. (2012a). Introduction of the alimentary tract with its microbes. In: Hakalehto, E. (ed.). Alimentary Microbiome – A PMEU Approach. New York, NY, USA: Nova Science Publishers, Inc.

Hakalehto, E. (2012b). Research on the normal microflora and simulations of its interactions with the PMEU. In: Hakalehto, E. (ed.). Alimentary Microbiome – A PMEU Approach. New York, NY, USA: Nova Science Publishers, Inc.

Hakalehto, E. (2013a). Interactions of *Klebsiella* sp. with other intestinal flora. In: Pereira, L. A., Santos, A. (ed.). *Klebsiella* Infections: Epidemiology, Pathogenesis and Clinical Outcomes. New York, USA: Nova Science Publishers, Inc., 1–33.

Hakalehto, E. (2013b). Enhanced mycobacterial diagnostics in liquid medium by microaerobic bubble flow in Portable Microbe Enrichment Unit. Pathophysiology, 20: 177–180.

Hakalehto, E. (2015a). Microbes and human digestive system. In: Hakalehto, E. (ed.). Microbiological Clinical Hygiene. New York, NY, USA: Nova Science Publishers, Inc., 219–258.

Hakalehto, E. (2015b). Microbiological Food Hygiene. New York, NY, USA: Nova Science Publishers, Inc., 2015.

Hakalehto, E. (2015c). Hazards and prevention of food spoilage. In: Hakalehto, E. (ed.). Microbiological Food Hygiene. New York, NY, USA: Nova Science Publishers, Inc.

Hakalehto, E. (2015d). Mycobacterial detection views. In: Hakalehto, E. (ed.). Microbiological Clinical Hygiene. New York, NY, USA: Nova Science Publishers, Inc.

Hakalehto, E. (2015e). Hygienic lessons from the dairy microbiology cases. In: Hakalehto, E. (ed.). Microbiological Food Hygiene. New York, NY, USA: Nova Science Publishers, Inc.

Hakalehto, E. (2015f). Antibiotic resistance in foods. In: Hakalehto, E. (ed.). Microbiological Food Hygiene. New York, NY, USA: Nova Science Publishers, Inc. pp. 233–251.

Hakalehto, E. (2016). The many microbiomes. In: Hakalehto, E. (ed.). Microbiological Industrial Hygiene. New York, NY, USA: Nova Science Publishers, Inc.

Hakalehto, E. (2018). Modes and consequences of the microbial interactions. In: Hakalehto, E. (ed.). Microbiological Environmental Hygiene. New York, NY, USA: Nova Science Publishers, Inc.

Hakalehto, E. (2020). Current megatrends in food production related to microbes. 5th International Conference on Food Chemistry and Technology (FCT-2019), Los Angeles, USA, 4.-6.11.2019. Journal of Food Chemistry and Nanotechnology, 6 (1): 78–87.

Hakalehto, E. (2021a). Chicken IgY antibodies provide mucosal barrier against SARS-CoV-2 virus and other pathogens. IMAJ, 23: 208–211.

Hakalehto, E. (2021b). Patient examples of imbalanced microbiome with antibiotic, toxic or metabolic emissions. Lecture. 12th International Congress on Autoimmunity – Virtual Congress.

Hakalehto, E. (2021c). Probiotics and prebiotics balance the food uptake and gut defences. 7th International Conference Food Chemistry & Technology (FCT-2021), scheduled for November 08–10, 2021; Paris, France.

Hakalehto, E. (2021d). Probiotics, antibiotics and politics. In: UVC-LED BLOG & NEWS. LED FUTURE. First published on 6th of April, 2021. Modified from the original blog article.

Hakalehto, E., Finne, J. (1990). Identification by immunoblot analysis of major antigenic determinants of the anaerobic beer spoilage bacterium genus *Pectinatus*. FEMS Microbiology Letters 67 (3), 307–311, https://doi.org/10.1111/j.1574-6968.1990.tb04038.x

Hakalehto, E., Hänninen, O. (2012). Gaseous CO_2 signal initiate growth of butyric acid producing *Clostridium* butyricum both in pure culture and in mixed cultures with *Lactobacillus brevis*. Canadian Journal of Microbiology, 58: 928–931. doi: http://dx.doi.org/10.1139/w2012-059.

Hakalehto, E., Heitto, L. (2012). Minute microbial levels detection in water samples by Portable Microbe Enrichment Unit technology. Environment and Natural Resources Research, 2, 80–88.

Hakalehto, E., Humppi, T. (2018). Monitoring microbes in the ambient air. In: Hakalehto, E. (ed.). Microbiological Environmental Hygiene. New York, NY, USA: Nova Science Publishers, Inc. pp. 383–404.

Hakalehto, E., Jaakkola, K. (2013). Synergistic effect of probiotics and prebiotic flax product on intestinal bacterial balance, Clinical Nutrition. 32, Supplement 1, S200.

Hakalehto, E., Haikara, A., Enari, T.-M., Lounatmaa, K. (1984). Hydrochloric acid extractable protein patterns of *Pectinatus cerevisiophilus* strains. Food Microbiology, 1(3), 209–216. https://doi.org/10.1016/0740-0020(84)90036-4

Hakalehto, E., Pesola, J., Heitto, L., Närvänen, A., Heitto, A. (2007). Aerobic and anaerobic growth modes and expression of type 1 fimbriae in *Salmonella*. Pathophysiology, 14, 61–69. http://dx.doi.org/10.1016/j.pathophys.2007.01.003

Hakalehto, E., Humppi, T., Paakkanen, H. (2008). Dualistic acidic and neutral glucose fermentation balance in small intestine: Simulation *in vitro*. Pathophysiology, 15, 211–220. http://dx.doi. org/10.1016/j.pathophys.2008.07.001

Hakalehto, E., Pesola, J., Heitto, A., Bhanj Deo B., Rissanen, K., Sankilampi, U., Humppi, T., Paakkanen, H. (2009). Fast detection of bacterial growth by using Portable Microbe Enrichment Unit (PMEU) and ChemPro100i® gas sensor. Pathophysiology, 16, 57–62. http:// dx.doi.org/10.1016/j.pathophys.2009.03.001

Hakalehto, E., Hell, M., Bernhofer, C., Heitto, A., Pesola, J., Humppi, T., Paakkanen, H. (2010). Growth and gaseous emissions of pure and mixed small intestinal bacterial cultures: Effects of bile and vancomycin. Pathophysiology, 17, 45–53. http://dx.doi.org/10.1016/j.pathophys.2009.07.003

Hakalehto, E., Heitto, A., Heitto, L., Humppi, T., Rissanen, K., Jääskeläinen, A., Hänninen, O. (2011). Fast monitoring of water distribution system with portable enrichment unit –Measurement of volatile compounds of coliforms and *Salmonella* sp. in tap water. Journal of Toxicology and Environmental Health Sciences, 3: 223–233.

Hakalehto, E., Heitto, A., Heitto, L. (2013a). Fast coliform detection in portable microbe enrichment unit (PMEU) with Colilert(®) medium and bubbling. Pathophysiology. 2013 Sep;20 (4): 257–262. doi: 10.1016/j.pathophys.2013.05.001. Epub 2013 Jun 20.

Hakalehto, E., Paakkanen, H., Heitto, A., King, K., Armon, R. (2013b). Rapid detection of food (milk) microbial contamination based on volatiles emission (as food spoilage indicators) detected by the Portable Microbe Enrichment Unit (PMEU). Presentation in ICEI 2013 (International Conference on Environmental Indicators), Trier, Germany; 2013.

Hakalehto, E., Nyholm, O., Bonkoungou, I. J., Kagambega, A., Rissanen, K., Heitto, A., Barro, N., Haukka, K. (2014a). Development of microbiological field methodology for water and food-chain hygiene analysis of *Campylobacter* spp. and *Yersinia* spp. in Burkina Faso, West Africa. Pathophysiology. 21(3): 219–229. doi: 10.1016/j.pathophys.2014.07.004. Epub 2014 Aug 4. PubMed PMID: 25156815.

Hakalehto, E., Heitto, A., Heitto, L., Rissanen, K., Pesola, I., Pesola, J. (2014b). Enhanced recovery, enrichment and detection of Mycobacterium marinum with the Portable Microbe Enrichment Unit (PMEU). Pathophysiology, 21: 231–235. doi: 10.1016/j.pathophys.2014.07.005. Epub 2014 Aug 4.

Hakalehto, E., Pesola, J., Heitto, A., Hänninen, H., Hendolin, P., Hänninen, O., Armon, R., Humppi, T., Paakkanen, H. (2015a). First detection of *Salmonella* contaminations. In: Hakalehto, E. (ed.). Microbiological Food Hygiene. New York, NY, USA: Nova Science Publishers, Inc. pp. 131–154.

Hakalehto, E., Heitto, A., Hyyrynen, O., Jokelainen, J., Jääskeläinen, A., Heitto, L. (2015b). Testing of coliformic bacteria for water departments using automated sampling and analyses. In: McCoy, G. (ed.). Coliforms: Occurrence, Detection Methods and Environmental Impact. New York, NY, USA: Nova Science Publishers Inc.

Hakalehto, E., Heitto, L., Heitto, A. (2016a). Hygiene control cases of milk, juices, beverages and raw water production. In: Hakalehto, E. (ed.). Microbiological Industrial Hygiene. New York, NY, USA: Nova Science Publishers, Inc., 121–131.

Hakalehto, E., Heitto, A., Kivelä, J., Laatikainen, R. (2016b). Meat industry hygiene, outlines of safety and material recycling by biotechnological means. In: Hakalehto, E. (ed.). Microbiological Industrial Hygiene. New York, NY, USA: Nova Science Publishers, Inc.

Hakalehto, E., Heitto, A., Hirvonen, I., Väätäinen, U. (2019a). Effect of UV-radiation on pathogens in situ and in wound infections. Poster. 8th Congress of European Microbiologists FEMS-2019, 7–11 July 2019, Glasgow, Scotland.

Hakalehto, E., Heitto, A., Immonen, A., Pesola, J., Hirvonen, I., Väätäinen, U. (2019b). UVC light prevents infection and autoimmune reaction caused by microbial contamination and antigen exposure – simulation studies with eukaryotic tissues and prokaryotic cells. Poster. 12th International Congress on Autoimmunity – Virtual Congress.

Hakalehto, E., Heitto, A., Väätäinen, U. (2021). Screening of antibiotic resistant strains in two Finnish private hospitals. Manuscript in preparation.

Hament, J.-M., Kimpen, J. L. L., Fleer, A., Wolfs, T. F. W. (1999). Respiratory viral infection predisposing for bacterial disease: A concise review. FEMS Immunology and Medical Microbiology, 26 (3–4): 189–195. doi: https://doi.org/10.1111/j.1574-695X.1999.tb01389.x.

Hauser, S. C., Pardi, D. S., Poterucha, J. J. (eds.). (2006). Mayo Clinic Gastroenterology and Hepatology Board Review, 2nd Edition. Mayo Clinic Scientific Press. Taylor & Francis Group.

Havenaar, R. (2011). Intestinal health functions of colonic microbial metabolites: A review. Beneficial Microbes, 2 (2): 103–114. doi: 10.3920/BM2011.0003.

Hayashi, M. A., Bizerra, F. C., Da Silva, P. I. Jr (2013). Antimicrobial compounds from natural sources. Frontiers in Microbiology, 15 July 2013. doi: https://doi.org/10.3389/fmicb.2013.00195.

Helander, I., Hakalehto, E., Ahvenainen, J., Haikara, A. (1983) Characterization of lipopolysaccharides of *Pectinatus cerevisiophilus*. FEMS Microbiology Letters. 18 (3) 223-226. https://doi.org/10.1111/j.1574-6968.1983.tb00482.x

Helander, I., Saukkonen, K., Hakalehto, E., Vaara, M. (1984). Biological activities of lipopolysaccharides from *Pectinatus cerevisiophilus*. FEMS Microbiology Letters. 24 (1), 39–42. https://doi.org/10.1111/j.1574-6968.1984.tb01240.x

Hell, M., Bernhofer, C., Pesola, J., Pesola, I., Hakalehto, E. (2015). Prevalence, detection and prevention of foodborne outbreaks related to large hospital kitchens. In: Hakalehto, E. (ed.) Microbiological Food Hygiene. New York, NY, USA: Nova Science Publishers, Inc. pp. 91–110.

Hokajärvi, A.-M., Rytkönen, A., Tiwari, A., Kauppinen, A., Oikarinen, S., Lehto, K.-M., Kankaanpää, A., Gunnar, T., Al-Hello, H., Blomqvist, S., Miettinen, I. T., Savolainen-Kopra, C., Pitkänen, T. (2021). The detection and stability or the SARS-CoV-2 RNA biomarkers in wastewater influent in Helsinki, Finland. The Science of the Total Environment, 770: 145274. doi: 10.1616/j.scitotenv.2021.145274.

Holmes, E., Wist, J., Masuda, R., Lodge, S., Nitschke, P., Kimhofer, T., Loo, R. L., Begum, S., Boughton, B., Yang, R., Morillon, A. C., Chin, S. T., Hall, D., Ryan, M., Bong, S. H., Gay, M., Edgar, D. W., Lindon, J. C., Richards, T., Yeap, B. B., Pettersson, S., Spraul, M., Schaefer, H., Lawler, N. G., Gray, N., Whiley, L., Nicholson, J. K. (2021). Incomplete systemic recovery and metabolic phenoreversion in post-acute-phase nonhospitalized COVID-19 patients: Implications for assessment of post-acute COVID-19 syndrome. Journal of Proteome Research, 20 (6): 3315–3329. doi: 10.1021/acs.jproteome.1c00224.

Hristov, N. I., Betke, M., Theriault, D. E. H., Bagchi, A., Kunz, T. H. (2010). Seasonal variation in colony size of Brazilian free-tailed bats at Carlsbad Cavern based on thermal imaging. Journal of Mammalogy, 91: 183–193.

Humppi, T., Mustalahti, S., Lehto, T., Hakalehto, E. (2016). In situ decontamination of airborne *Bacillus atrophaeus* spores by vaporized hydrogen peroxide (VHP). In: Hakalehto, E. (ed.). Microbiological Industrial Hygiene. New York, NY, USA: Nova Science Publishers, Inc.; pp. 93–101.

Jaenisch, R., Bird, A. (2003). Epigenetic regulation of gene expression: How the genome integrates intrinsic and environmental signals. Nature Genetics, 33: 245–254. doi: https://doi.org/10.1038/ng1089.

Jahkola, M., Lindström, M., Korkeala, H. (1999). Botulismi EU-maissa. (In Finnish). Kansanterveys, 4: 7–8.

Jay, J. M. (1986) Modern Food Microbiology, Third Edition. New York, USA: Van Nostrand Reinhold.

Kojima, S., Kamada, A. J., Parrish, N. F. (2021). Virus-derived variation in diverse human genomes. Plos Genetics. April 26, 2021. doi: https://doi.org/10.1371/journal.pgen.1009324.

Konto-Ghiorghi, Y., Mairey, E., Mallet, A., Dumenil, G., Caliot, E., Trieu-Cuot, P., Dramsi, S. (2009). Dual role for pilus in adherence to epithelial cells and biofilm formation in *Streptococcus agalactiae*. J Plos Org. doi: doi.org/10.1371/journal.ppat.1000422.

Kuosmanen, T., Torvinen, A., Humppi, T., Hakalehto, E. (1995). Luminometric detection method for *Salmonella typhi*. Proceedings of the 5th International Symposium: Protection Against Chemical and Biological Warfare Agents, Stockholm, Sweden, 11–16 June 1995. Suppl., p. 238.

Laatikainen, R., Laatikainen, P., Martonen, H., Tiainen, M., Hakalehto, E. (2016). Quantitative quantum mechanical NMR analysis: The superior tool for analysis of biofluids. IECM 1st International Electronic Conference on Metabolomics.

Laitiomäki, E., Pesola, J., Hakalehto, E. (2015). Projected improvement in the fast microbiological analysis of neonatal blood samples. Journal of Neonatal Nursing, 21: 58–62.

Larsson, A., Carlander, D. (2003). Oral immunotherapy with yolk antibodies to prevent infections in humans and animals. Upsala Journal of Medical Sciences, 108 (2): 129–140. PMID: 14649324.

Lee, S. Y., Mabee, M. S., Jangaard, N. O. (1978). *Pectinatus*, a new genus of the family *Bacteroidaceae*. International Journal of Systematic Bacteriology. 28: 582–594.

Lee, S. Y., Moore, S. E., Mabee, M. S. (1981). Selective-differential medium for isolation and differentiation of *Pectinatus* from other brewery microorganisms. Applied and Environmental Microbiology, 41 (2): 386–387. doi: 10.1128/aem.41.2.386-387.1981.

Leyer, G. J., Johnson, E. A. (1990). Repression of toxin production by tryptophan in *Clostridium botulinum* type E. Archives of Microbiology, 154: 443–447.

Low, C. X., Tan, L. T.-H., Mutalib, N.-S., A. et al. (2021). Unveiling the impact of antibiotics and alternative methods for animal husbandry: A Review. Antibiotics (Basel) May 13, 10 (5): 578. doi: 10.3390/antibiotics10050578.

Lucia Rocha Carvalho, M., Beninga Morais, T., Ferraz Amaral, D., Maria Sigulem, D. (2000). Hazard analysis and critical control point system approach in the evaluation of environmental and procedural sources of contamination of enteral feedings in three hospitals. JPEN Journal of Parenteral and Enteral Nutrition, 24: 296–303. doi: 10.1177/0148607100024005296. PMID: 1111785.

Macnab, R. M. (2004). Type III flagellar protein export and flagellar assembly. Biochimica et Biophysica Acta. 2004 Nov 11; 1694 (1–3): 207–217. doi: 10.1016/j.bbamcr.2004.04.005. PMID: 15546667.

Maisey, E. A., Wadsworth, J. D., Poulain, B., Shone, C. C., Melling, J., Gibbs, P., Tauc, L., Dolly, J. O. (1988). Involvement of the constituent chains of botulinum neurotoxins A and B in the blockade of neurotransmitter release. European Journal of Biochemistry / FEBS, 177: 683–691.

Marzoog, B. A., Vlasova, T. I. (2021). The possible puzzles of BCG vaccine in protection against COVID-19 infection. The Egyptian Journal of Bronchology, 15 (1): 7. doi: https://doi.org/10.1186/s43168-021-00052-3.

Mawdsley, S. (2020). What are acute NHS trusts in England doing to prevent the cross-border spread of carbapenem-resistant *Enterobacteriaceae*? Journal of Infection Prevention, 21 (5): 196–201. doi: 10.1177/1757177420935633. Epub 2020 Jul 21. PMID: 33193822; PMCID: PMC7607406.

Mayo Clinic (2020). Patient Care & Health Information. Diseases and Conditions. Botulism. Mayo Foundation for Medical Education and Research (MFMER). Published on 12 August 2020. https://www.mayoclinic.org/diseases-conditions/botulism/diagnosis-treatment/drc-20370266.

Mentu, J. V., Heitto, L., Keitel, H. V., Hakalehto, E. (2009). Rapid microbiological control of paper machines with PMEU method. Paperi ja Puu / Paper and Timber, 91, 7–8 (90th Anniversary n:o).

Mukherjee, S. (2020). Before virus, after virus: A reckoning. Cell. 15; 183 (2): 308–314. doi: 10.1016/j.cell.2020.09.042. PMID: 33064987; PMCID: PMC7560376.

Mullis, K. F., Faloona, F., Scharf, S., Saiki, R., Horn, G., Erlich, H. (1986). "Specific enzymatic amplification of DNA in vitro: The polymerase chain reaction". Cold Spring Harbor Symposia on Quantitative Biology, 51: 263–273. doi: 10.1101/sqb.1986.051.01.032. PMID 3472723.

Nakkarach, A., Foo, H. L., Song, A. A., Mutalib, N. E. A., Nitisinprasert, S., Withayagiat, U. (2021). Anti-cancer and anti-inflammatory effects elicited by short chain fatty acids produced by *Escherichia coli* isolated from healthy human gut microbiota. Microbial Cell Factories Feb 5; 20 (1): 36.doi: 10.1186/s12934-020-01477-z.

Nguyen, H. H., Tumpey, T. M., Park, H. J., Byun, Y. H., Tran, L. D., Nguyen, V. D., Kilgore, P. E., Czerkinsky, C., Katz, J. M., Seong, B. L., Song, J. M., Kim, Y. B., Do, H. T., Nguyen, T., Nguyen, C. V. (2010). Prophylactic and therapeutic efficacy of avian antibodies against influenza virus H5N1 and H1N1 in mice. PLoS One. 2010 Apr 13;5 (4): e10152. doi: 10.1371/journal.pone.0010152. PMID: 20405007; PMCID: PMC2854139.

Niewiesk, S. (2014). Maternal antibodies: Clinical significance, mechanism of interference with immune responses, and possible vaccination strategies. Frontiers in Immunology, 5: 446. doi: 10.3389/fimmu.2014.00446.

Oxford, J. S., Gill, D. (2018). Unanswered questions about the 1918 influenza pandemic: origin, pathology, and the virus itself. Lancet Infectious Diseases. 18 (11):e348–e354. doi: 10.1016/S1473-3099(18)30359-1. Epub 2018 Jun 20. PMID: 29935779.

Pellizzari, R., Rossetto, O., Washbourne, P., Tonello, F., Nicotera, P. L., Montecucco, C. (1998). In vitro biological activity and toxicity of tetanus and botulinum neurotoxins. Toxicology Letters, 102–103: 191–197.

Pérez de la Lastra, J. M., Baca-González, V., Asensio-Calavia, P., González-Acosta, S., Morales-delanuez, A. (2020). Can immunization of hens provide oral-based therapeutics against COVID-19? Vaccines (Basel). 2020 Aug 28; 8 (3): 486. doi: 10.3390/vaccines8030486. PMID: 32872186; PMCID: PMC7565424.

Pesola, J., Hakalehto, E. (2011). Enterobacterial microflora in infancy – a case study with enhanced enrichment. Indian Journal of Pediatrics, 78: 562–568. doi: http://dx.doi.org/10.1007/s12098-010-0341-5.

Pesola, J., Vaarala, O., Heitto, A., Hakalehto, E. (2009). Use of portable enrichment unit in rapid characterization of infantile intestinal enterobacterial microbiota. Microbial Ecology in Health and Disease, 21, 203–210. http://dx.doi.org/10.3109/08910600903367810.

Pesola, J., Paakkanen, H., Hakalehto, E. (2012). Enhanced diagnostics of pyelonephritis – a case study. International Journal of Medicine and Medical Sciences, 2: 273–277.

Pesola, J., Humppi, T., Hakalehto, E. (2015). Method development for *Clostridium botulinum* toxin detection. In: Hakalehto, E. (ed.). Microbiological Food Hygiene. New York, NY, USA: Nova Science Publishers, Inc, 197–209.

Petrosillo, N., Viceconte, G., Ergonul, O., Ippolito, G., Petersen, E. (2020). COVID-19, SARS and MERS: Are they closely related? Clinical microbiology and infection: The Official Publication of the European Society of Clinical Microbiology and Infectious Diseases. 2020 Jun; 26 (6): 729–734. doi: 10.1016/j.cmi.2020.03.026. Epub 2020 Mar 28. PMID: 32234451; PMCID: PMC7176926.

Pitkänen, T., Bräcker, J., Miettinen, I., Heitto, A., Pesola, J., Hakalehto, E. (2009). Enhanced enrichment and detection of thermotolerant *Campylobacter* species from water using the Portable Microbe Enrichment Unit (PMEU) and realtime PCR. Canadian Journal of Microbiology, 55: 849–858.

Platto, S., Xue, T., Carafoli, E. (2020). COVID19: an announced pandemic. Cell Death and Disease. 11, 799. https://doi.org/10.1038/s41419-020-02995-9.

Proksch, G., Baganz, D. (2020). CITYFOOD: Research Design for an International, Transdisciplinary Collaboration. Technology|Architecture + Design, 4: 1, 35–43. doi: 10.1080/24751448.2020.1705714.

Quammen, D. (2014). Ebola: The natural and human history of a deadly virus. New York, USA: W. W. Norton & Co., Inc.

Rohde, M. (2019). The gram-positive bacterial cell wall. Microbiology Spectrum, 7 (3). doi: 10.1128/microbiolspec.GPP3-0044-2018.

Ruiz-Palacios, G. M., Torres, N., Ruiz-Palacios, B. R., Torres, J., Escamilla, E., Tamayo, J. (1983). Cholera-like enterotoxin produced by *Campylobacter jejuni*. Lancet, 322 (8344): 250–253. doi:https://doi.org/10.1016/S0140-6736(8390234-9).

Sachi, S., Ferdous, J., Sikder, M. H., Azizul, K., Hussani, S. M. (2019). Antibiotic residues in milk: Past, present, and future. Journal of Advanced Veterinary and Animal Research. 2019; 6 (3): 315–332. Published 2019 Jul 11. doi: 10.5455/javar.2019.f3500.

Samgina, T., Vorontsov, E., Gorshkov, V., Hakalehto, E., Hänninen, O., Zubarev, R., Lebedev, A. (2012). Composition and antimicrobial activity of the skin peptidome of Russian brown frog *Rana temporaria*. Journal of Proteome Research, 11 (12): 6213–6222.

Samgina, T. Y., Tolpina, M. I., Hakalehto, E., Artemenko, K. A., Bergquist, J., Lebedev, A. T. (2016). Proteolytic degradation and deactivation of amphibian skin peptides obtained by electrical stimulation of their dorsal glands. Analytical and Bioanalytical Chemistry, 408: 3761–3768.

Sathyamoorthy, V., Dasgupta, B. R., Foley, J., Niece, R. L. (1988). Botulinum neurotoxin type A: Cleavage of the heavy chain into two halves and their partial sequences. Archives of Biochemistry and Biophysics, 266: 142–151.

Schleifer, K. H., Leuteritz, M., Weiss, N., Ludwig, W., Kirchhof, G., Seidel-Rüfer, H. (1990). Taxonomic study of anaerobic, gram-negative, rod-shaped bacteria from breweries: Emended description of *Pectinatus cerevisiiphilus* and description of *Pectinatus frisingensis* sp. nov., *Selenomonas lacticifex* sp. nov., *Zymophilus raffinosivorans* gen. nov., sp. nov., and *Zymophilus paucivorans* sp. nov. International Journal of Systematic Bacteriology, 40 (1): 19–27. doi: 10.1099/00207713-40-1-19.

Si, F., Le Treut, G., Sauls, J. T., Vadia, S., Levin, P. A., Jun, S. (2019). Mechanistic origin of cell-size control and homeostasis in bacteria. Current Biology: CB Jun 3; 29 (11): 1760–1770.e7. doi: 10.1016/j.cub.2019.04.062. Epub 2019 May 16.

Silva, O. N., Mulder, K. C., Barbosa, A. E., Otero-Gonzalez, A. J., Lopez-Abarrategui, C., Rezende, T. M., Dias, S. C., Franco, O. L. (2011). Exploring the pharmacological potential of promiscuous host-defense peptides: From natural screenings to biotechnological applications. Published on 22 November 2011. Frontiers in Microbiology. doi: 10.3389/fmicb.2011.00232.

Sperber, W. H., Stier, R. F. (2009). Happy 50th birthday to HACCP: Retrospective and prospective. FoodSafety Magazine: 42–46. Retrieved 11 January 2015.

Suriyasathaporn, W., Chupia, V., Sing-Lah, T., Wongsawan, K., Mektrirat, R., Chaisri, W. (2012). Increases of antibiotic resistance in excessive use of antibiotics in smallholder dairy farms in northern Thailand. Asian-Australasian Journal of Animal Sciences, 25 (9): 1322–1328. doi: 10.5713/ajas.2012.12023.

Van de Perre, P. (2003). Transfer of antibody via mother's milk. Vaccine. 2003 Jul 28; 21 (24): 3374–3376. doi: 10.1016/s0264-410x(03)00336-0. PMID: 12850343.

Wang, H., Wang, J., Li, S., Li, J., Jing, C. (2019). Prevalence of antibiotic resistance genes in cell culture liquid waste and the virulence assess for isolated resistant strains. Environmental Science and Pollution Research, 26 (31): 32040–32049.doi: 10.1007/s11356-019-06299-0. Epub 2019 Sep 6.

Velázquez, O. C., Lederer, H. M., Rombeau, J. L. (1997). Butyrate and the colonocyte. Production, absorption, metabolism, and therapeutic implications. Advances in Experimental Medicine and Biology, 427: 123–134.

Williams, S. (2020). A brief history of human coronaviruses. The Scientist, June 2.

Wirtanen, G., Salo, S. (2010). 41st R^3-Nordic Symposium: Cleanroom technology, contamination control and cleaning: Dipoli, Espoo, Finland, May 25–26, 2010.

Wong, N. A., Saier, M. H. Jr (2021). The SARS-coronavirus infection cycle: A survey of viral membrane proteins, their functional interactions and pathogenesis. International Journal of Molecular Sciences, 22 (3): 1308. doi: https://doi.org/10.3390/ijms22031308

Östergaard, L. (2021). SARS-CoV-2 related microvascular damage and symptoms during and after COVID-19: Consequences of capillary transit-time changes, tissue hypoxia and inflammation. Physiological Reports. 2021 Feb; 9 (3): e14726. doi: 10.14814/phy2.14726.

Jukka M. Sauramäki, Frank Adusei-Mensah
Jukka-Pekka Hakalehto, Robert Armon, Elias Hakalehto

7 Pandemic situation and safe transportation, storage, and distribution for food catering and deliveries

Abstract: How do you get your food? This is one of the basic issues in life for all of us. The logistics of food deliveries has changed during different periods. The world population has urbanized in ever-increasing numbers, which has challenged our digestive and overall health. The meals in general have become less versatile and increasingly nutrient poor (qualitative malnutrition). The modern lifestyle and the one-sided stress on our autonomic nervous system has caused more and more anxiety and depression in the cities. Consequently, the smoothness of food deliveries always is an important factor in mental and physical well-being of citizens.

The pandemic conditions have made the stress still more universal also among the food-chain personnel. Novel solutions have been implemented in urgent manner for the protection of the customers and workers alike. Several already existing technologies should have been and will be implemented with extreme urgency. As most of the societies were caught unprepared for the COVID-19 pandemics, this caused extra strain also for the food distribution sector, which had to adapt quickly into the entirely new working conditions and business models.

The lack of microbiological experience in the companies, municipalities, social systems, and welfare services, and even in the healthcare sector was most tangible during the onset of the current crisis. Since the SARS-CoV-2 viruses rapidly spread all over the world, it posed additional requirement for the food services. Restaurants were closed at least partially, and both the meals at work and at home needed extra caution. The professionals needed rapid education for dealing with the pandemics, as well as more cooperation with the authorities and healthcare specialists.

Moreover, the exceptional times and practices initiated several unpredictable issues also microbiologically. For example, the infections caused by *Listeria* grew fast in numbers. These previously known, but in increased numbers occurring hygienic issues have caused extra concern.

Jukka M. Sauramäki, Posti Oy, Helsinki, Finland
Frank Adusei-Mensah, Finnoflag Oy, Kuopio and Siilinjärvi, Finland; Institute of Public Health and Clinical Nutrition, University of Eastern Finland, Kuopio, Finland
Jukka-Pekka Hakalehto, Finnoflag Oy, Kuopio and Siilinjärvi, Institute of Dentistry, University of Eastern Finland
Robert Armon, Technion, Israel University of Technology, Haifa, Israel
Elias Hakalehto, Finnoflag Oy, Kuopio and Siilinjärvi, Finland; Department of Agricultural Sciences, University of Helsinki, Helsinki, Finland; University of Eastern Finland, Kuopio, Finland

https://doi.org/10.1515/9783110724967-008

Regular fast-food restaurants also had to reform their services. For example, the drive-in grills and takeaway desks needed novel applications and equipment. Such methods could be supported by sophisticated means like novel ventilation techniques, disinfectants, gases, or ultraviolet C light.

Since the food distribution service is a part of the society, and it operates among the surrounding population, a clear view of the epidemic situation is necessary background information for the daily function of it. Then an updated view of the levels of disease-causing agents could be obtained by new rapid microbiology methods, or by screening of the viruses or other germs from the wastewater. It is noteworthy, that any worsening of the epidemic situation could influence radically on the health risks, as well as on the precautionary measures.

7.1 Introduction

During the pandemic era, the overall control of customer contacts should be carefully planned. It is important that no transmission of disease-causing agents is occurring via food or human-to-human contacts. Specific equipment for the safe food delivery should be designed.

After the first steps in the eradication of the viral diseases, it is still needed to maintain extra caution. The individuals dependent on the food distribution and home delivery services are often the most vulnerable ones to the infections. Some future (and in some cases also present-day) technologies include transportation drones, which could be used for food deliveries (Figure 7.1).

The equipment and premises could be disinfected by gaseous disinfectants, such as hydrogen peroxide (Lineback et al., 2018). In case of atmospheric sampling of the outdoor microbes, we have used formaldehyde gasification for the sterilization of the sample collectors (Casella and Schmidt-Lorenz, 1989). However, formaldehyde is not necessarily 100% safe for such purposes in commercial solutions. It would be important to test various strategies for implementing clean food transport vehicles, and for preventing any contagion during the deliveries. The various pre-pandemic practices are described and discussed by Sauramäki and Hakalehto (2015). Also, air traffic and transportation should be carefully monitored as it is potentially distributing microbes very fast internationally.

One potential technology would be the use of ultraviolet C (UVC) disinfection of the surfaces of food packages, for instance. This technology has been in use for sterilizing mobile phones, car keys, or other potent vectors for contamination, which are in the joint use in food delivery companies (Figure 7.2). At the airports, it is possible to clean the luggage surfaces by the UVC technique. The microbiological safety instructions for traffic systems are discussed more broadly by Hakalehto et al. (2018).

Figure 7.1: Drones have been piloted in many delivery companies. Some of the restaurant meals are already delivered by drones in many cities around the world. Photo: Posti Group Oyj.

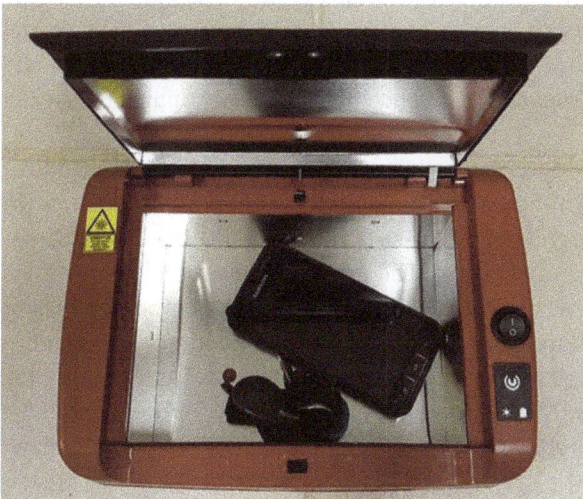

Figure 7.2: Handheld scanners and tracking devices, mobile phones, keys, and so on can be disinfected by UVC device (Led Future Oy, Kuopio, Finland). Photo: Jukka Sauramäki.

International coordinated measures against the COVID-19 dissemination were totally lacking or mostly on low gear when the COVID-19 pandemics occurred. However, in several densely populated Asian countries, more experience and hands on information on the prevention of contagious diseases had accumulated. Such "minor" pandemics as SARS-1 and swine flu had given more practical training for the authorities in those countries. Moreover, the general attitude of people was that they more prepared for the next round of the epidemics there. This was due to long history of epidemic diseases in the Asian subcontinent, sense of common or joint responsibility and other reasons, besides the more recent lessons on the viral pandemics (Majumder and Minko, 2021). As a result of the current viral pandemic episode caused by the SARS-CoV-2, we should learn more skills and techniques to protect our populations and their food from various future pandemics, as well as the common hygienic risks associated with these episodes. Besides, we need to continue the food deliveries in normal times, as the pandemics will calm down, also taking into account the lessons learnt during the exceptional times.

The core aspect in nearly all work required for eradicating the infectious diseases is the reasonable level of general hygiene. In practice, the cleanliness needed is related to the suspected pathogen in question. However, during the hazardous phase, when the epidemics is rampaging around, all measures are important if they could be implemented in time, and save human lives. In the conference Biotech Japan in 2014 in Tokyo, the Japanese government representatives introduced results of their extensive research (Figure 7.3). According to this data, the best possible means for improving public health is to lower the threshold for implementing novel devices and technologies for the battle against diseases, contagions, and contaminations. In our mind, the same common truth and strategy is most applicable for the food sector, too.

In this chapter, we briefly describe the history of reasons behind the pandemics in general. In fact, many of the epidemic agents are either (a) airborne, or (b) food or water transmittable, or (c) both. In case of the recent COVID-19, the contagious viral agent belonged to the group c. The pre-emptive action also includes the surface hygiene, as most surfaces are contaminated by air-borne particles or during contact with diseased individuals. Moreover, direct contact between humans has also been recognized as one of the main vehicles of disseminating the infections (Bahl et al., 2020). All the potential means of distribution need to be considered as possible routes of contaminations in case of municipal and food hygiene. Therefore, we look at the hygiene of the sewage systems since they are important sources for contamination in the cities. The food sector personnel are subjected to the same threats as the general public, which is important to be kept in mind.

The divergence of microbiological cleanliness or hygiene maintenance methods has been always quite the same and at all circumstances. However, during the

Figure 7.3: Finnoflag Oy team in Biotech Japan 2014. From left to right Jukka-Pekka and Elias Hakalehto, Jouni Pesola, Lauri and Matti Heitto. This important conference was participated by more than 10,000 delegates. The highlight lecture was given by Japanese Government representatives publishing the research data on the primary importance of relieving bureaucratic obstacles hindering the implementation of novel medical technologies. Photo: Aki Immonen.

pandemic periods, more emphasis is on the effectiveness of the methods. In fact if the baseline for microbiological safety could be achieved by:

1. personal protection (masks, hand disinfection, distancing, etc.) (Armon, 2015);
2. cleaning up the surfaces, handles, tools, telephones, computers, and other items (Hakalehto et al., 2015);
3. improved hygiene of the ambient air (including the restriction of aerosol formation) (Hakalehto and Humppi, 2018);
4. water hygiene of both household and recreational waters, as well as the municipal or industrial water sources (Hakalehto, 2014–2015);
5. high standard of wastewater purification with special emphasis on hospital wastewaters (Hakalehto and Väätäinen, 2016);
6. industrial hygiene of the food manufacturing, storage, distribution, and recycling chains (Hakalehto, 2015b, d);
7. hygiene control at farms and in the agriculture with special reference to zoonoses (Armon, 2015; Immonen et al., 2015).

The environmental health background has always depended on the readiness of the societies around us.

7.2 Changes in the societies and the revised estimates of risk situations

Already, prior to the current pandemics, the urbanization of global population in all countries has brought about challenges on individual and public health. For example, the hectic lifestyle in cities probably has caused the increasing IBD/IBS incidences in most countries (Hakalehto, 2020).

The coronavirus pandemic has changed the societies. It is not the first pandemics in the world history. Thanks to the modern medicine and vaccinations, the outcome of the infectious waves has not been as disastrous as it could have been, or as it has unfortunately been for so many times during the centuries.

More than 100 years ago, the so-called Spanish flu caused a terrible death toll, which has been estimated to have been around 50–100 million deceased patients worldwide. The current pandemic of COVID-19 has demanded less than 4 million lives up to June 2021. The hazardous condition, however, is not over yet. In several big countries, such as India or Brazil, people are still in a difficult situation. Although the vaccinations are given everywhere, the risks have not been fully overcome. For example, new viral variants may be less susceptible to the immune response obtained by vaccination or by going through the sickness. In an interview by BBC, the Imperial College virologist Wendy Barclay in June 2021 popularized the dissemination and modifications of a virus. The R_O number gives the average number of individuals one diseased person transmits the infection. If the number is more than 1, the disease is gaining more foothold in the population. The R_O of the delta-variant (originating from India) could be up to 8. This variant is 50% more infective than the so-called alfa-variant, which in turn is 50% more infective than the original viral strain of SARS-CoV-2 originally isolated in Wuhan, China.

Another aspect of the hazards caused by the virus epidemics, is the threat of bacterial or fungal complications. During the Spanish flu, 70% of the deaths were caused by pneumococcal, staphylococcal, and other bacterial infections following the influenza virus epidemics (Morens et al., 2008). The viral attack first damaged the epithelial cells and attenuated the immune protection; then the bacteria could cause an infection more readily in the weakened individuals. According to some data, five months after remission from COVID-19, 80% of the patients were still receiving the antibiotics (Huang et al., 2021).

In the aftermath of the virus, several so-called "long Covid" symptoms have been reported (Östergaard, 2021). For example, damage is not caused in the respiratory tract only, but such vital organs as heart, kidneys, and reproductive tissues are also at risk. Male infertility could be caused by the virus (Hallak et al., 2021; Tian and Zhou, 2021). Also, the overall condition of the slowly recovering patients has lowered in many cases. For some athletes, this has caused troubles for their careers.

Besides the health effects of COVID-19, this pandemic has pushed world economics into big troubles, for instance in Africa (Mahama, 2020).

Since the effect of pandemics influences the entire society, it also changes all practices and procedures in the food sector. Therefore, the understanding on the origins, development, safety measures, damage and monitoring of the microbiological agent is of highly relevance. At best, it could prove out to be the onset of new businesses. However, it is of crucial importance that the hygiene monitoring and control is prevailed on adequate levels, at all times.

7.3 General remarks – the "situation room"

When the COVID-19 pandemic started to influence the everyday life in various societies, it became important also for the companies operating on food sector to follow up the development of the general situation of the epidemics, such as the incidences, numbers of cases, aerial status, and the newest instructions (Figure 7.4). In Finland this information was officially produced by Finnish Institute for Health and Welfare, and actively reported by all news media. Internationally, the World Health Organization (WHO) has screened the global situation. Also, some big universities, such as Johns Hopkins in Baltimore, Maryland, USA, have kept records on the international course of the pandemics. The transparent publicity of the statistics on the epidemics is of paramount importance.

To follow and react according to these statistics has become a part of everyday operative routines in many companies and increased cooperation between employers and occupational health service providers. Situation rooms were founded in firms to receive this information and to instruct operative units to implement protection methods and other guidelines given by the authorities.

7.4 COVID-19 and sewage

The epidemiology of sewage surveillance for therapeutic drugs, pathogen screening, or steroid usage is a well-known and cost-effective practice. The principle behind SARS-CoV-2 sewage surveillance is the following: an infected person (both with symptoms and without symptoms) sheds the virus in his urine, feces, sputum, and other means from the time of infection, and the shedding can continue for several weeks before they are no longer infectious. Several countries including Finland, China, Netherlands, South Africa, Australia, and Brazil are monitoring SARS-CoV-2 viral shedding in sewage to track down numbers of cases, predict surges, plan targeted testing, and estimate prevalence (Kreier, 2021; Richardson, 2021). Through this approach Hong Kong researchers are monitoring sewage in apartment

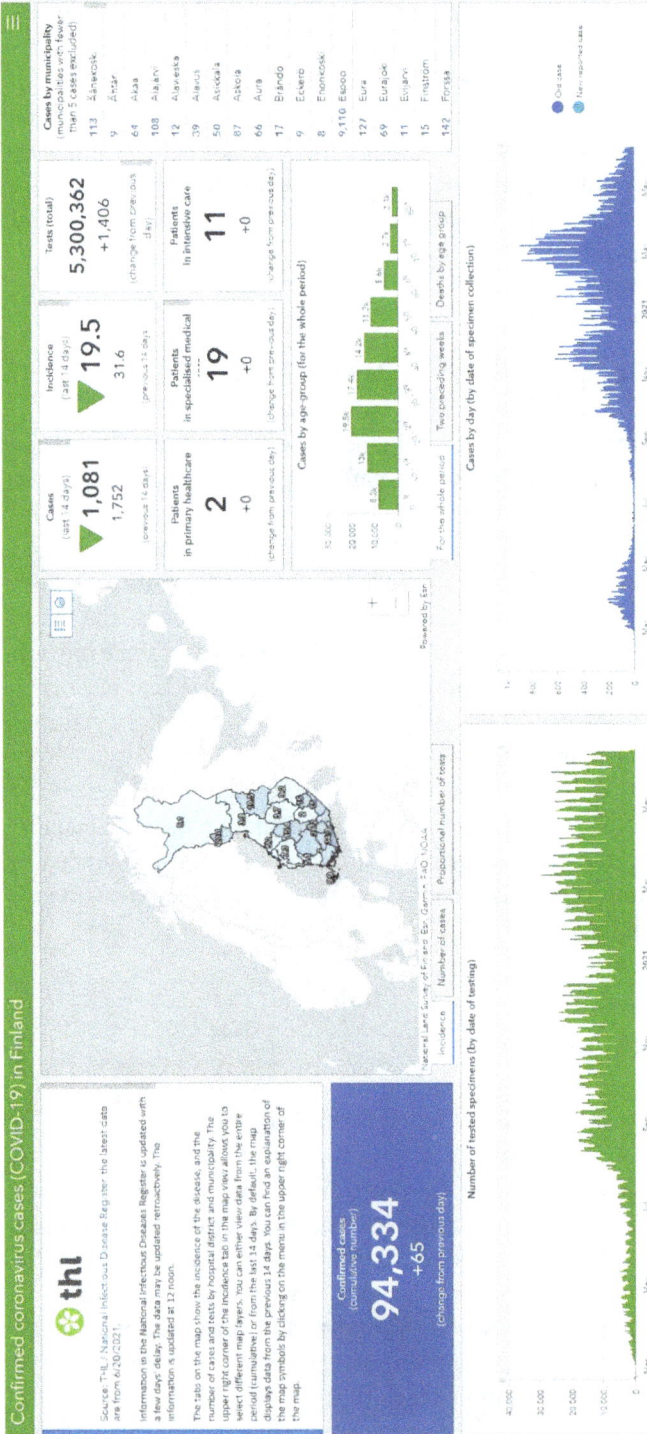

Figure 7.4: National virus dashboard by THL – Finnish Institute for Health and Welfare is widely used by organizational situation rooms in Finland.
Source: THL.

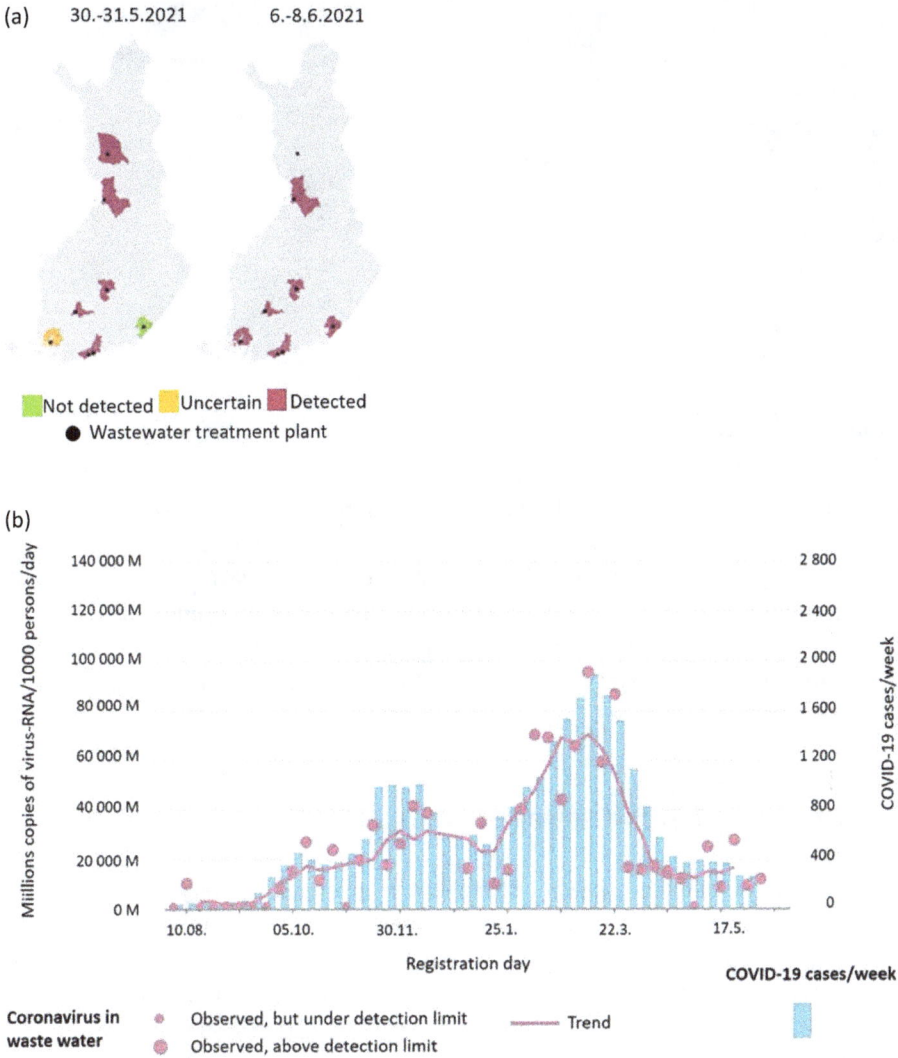

Figure 7.5: (a) Wastewater monitoring gives the situation room real-time data about the abundance of pathogens in various locations. SARS-CoV-2 RNA detection in wastewater at Finnish sewage farms. Source: THL (Finnish Institute for Health and Welfare, Expert Microbiology Unit, Kuopio, Finland). (b) The SARS-CoV-2 levels in sewage water in the Viikinmäki treatment plant between August 2020 and May 2021. Source: THL (Finnish Institute for Health and Welfare, Expert Microbiology Unit, Kuopio, Finland). (c) Sewage COVID-19 virus analysis. Surveillance results/ 100,000 citizens. The number of RNA copies is counted in every two weeks from all the participating municipal wastewater treatment plants. The sewage RNA results have corresponded to the outcome of COVID-19 in the communities (Hokajärvi et al, 2021). Source: THL (Finnish Institute for Health and Welfare, Expert Microbiology Unit, Kuopio, Finland).

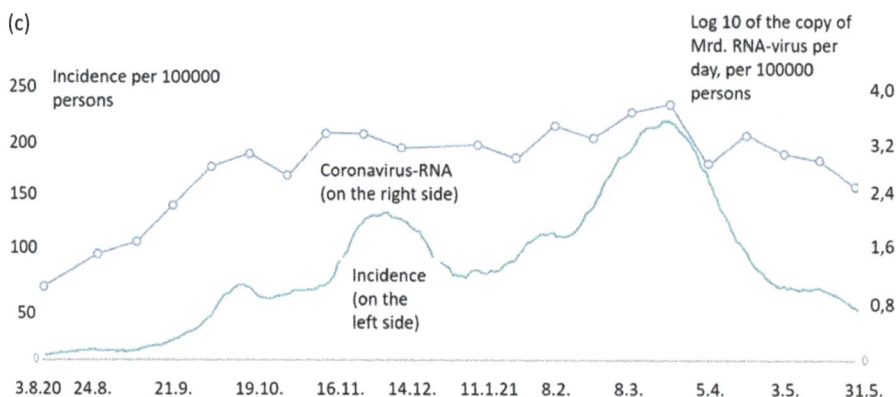

Figure 7.5 (continued)

buildings to find previously undetected infections. Moreover, viral variants entering into Yellowknife city were being monitored using this approach (Kreier, 2021).

In Finland, regular sewage monitoring of coronavirus levels have shown SARS-CoV-2 in the wastewater units of several cities including Viikinmäki in Helsinki, Suomenoja in Espoo, and Kakolanmäki in Turku. The wastewaters in the cities Tampere, Rauma and Vaasa have also contained coronavirus during August 2020 (Pitkänen, 2020; see Figure 7.5a–c).

7.4.1 Benefit of active sewage COVID-19 testing

Compared to mass testing, sewage surveillance approach is a very economical approach for monitoring the population. It safeguards privacy and helps in implementing correct constraints. It also serves as an early warning system to plan vaccination programs and to predict hospitalization numbers of days ahead of time (Kreier, 2021). It is a cheap and effective alternative for prediction with close to 85% success rate. Based on the report of the South China Morning Post, sewage surveillance system helped the health officials to find several asymptomatic cases in selected apartments with reduced cost (Kreier, 2021). In Netherlands, such testing has been helpful for the government officials to make policy decisions concerning lockdowns and COVID-19 resources allocations ahead of time. It is an objective tool to get a universal picture, though "Not everyone is getting tested, but everyone is going to the bathroom," according to Gertjan Medema (Kreier, 2021). This monitoring of virus (or bacterial situation) is very useful and informative, giving an idea of the likelihood of virus contamination in different areas, based on their abundance levels in the wastewater. This kind of monitoring approach could also give the food deliveries crucial information about the risk levels.

7.4.2 Could sewage water be a vehicle for SARS-CoV-2 infection?

In Guangzhou, China, it was reported that, contaminated sewage water might serve as a vehicle for possible transmission of COVID-19 outbreaks in communities (Yuan et al., 2020). The survival of the SARS-COV-2 viral particle has been investigated under different sewage conditions. In hospital sewer water, domestic sewage, and dechlorinated tap water, SARS-CoV-2 persisted for 2 days at 20 °C, and ≥14 days at 4 °C (Wang et al., 2005).

In vitro experiments showed that the COVID-19 virus survived 3 days in feces and 17 days in urine at 20 °C while SARS-CoV persistence in similar sewage types were higher ranging from 14 days in wastewater to at least 17 days in feces or urine at 4 °C (Wang et al., 2005). The human coronavirus has shown that the time required for the virus titer to decrease 99.9% (T99.9) corresponding to "inactivated" was 10 days at 23 °C in water and >100 days at 4 °C in water (Gundy et al., 2008). However, the inactivation (T99.9) was observed between 2 and 4 days for primary and secondary sludge in wastewater at 23 °C (Gundy et al., 2008).

It is important to plan the routes for food deliveries in the crowded city areas. In line with the previous knowledge, the coronavirus persistency in the wastewater is highly dependent on the temperature and wastewater treatment conditions. SARS-CoV-2 is more susceptible to disinfectants containing free chlorine than chlorine dioxide (Wang et al., 2005). In a systematic review, the conclusions were drawn, as follows (La Rosa et al., 2020):

a. SARS-CoV-2 have low stability in the environment at high temperatures and is sensitive to oxidants, like chlorine;
b. temperature is an important factor influencing viral survival; and
c. there is a lack of current evidence that human coronaviruses are present in surface or ground waters or are transmitted through contaminated drinking water (La Rosa et al., 2020).

7.4.3 Frozen food and COVID-19

Temperature variations affects both chemical and biochemical processes. Low-temperature exposures significantly prolong the lifespan of viruses. In the cold-chain food requiring refrigeration temperatures (4 °C), viruses containing coronavirus can persist from many days to several weeks (Chin et al., 2020). According to the WHO, coronaviruses are highly stable in frozen state, and could survive for up to 2 years at −20 °C (WHO, 2020). This emphasizes the importance of monitoring the frozen foods before the delivery. Also, the practices in the street corner sales of e.g. ice cream and other frozen or chilled refreshments should be adequately supervised (Figure 7.6).

Figure 7.6: Is there a return back to normal times with respect to unprotected queuing of ice cream? Any dairy products are at risk of bacterial contamination in addition to the risk of respiratory viruses being transmitted in the queue. Photo: Florencia Viadana on Unsplash.

To ensure the quality of foods, the temperatures for packaging, transportation, and storing the chilled and frozen foods during transportation operations were regulated below 8 °C for refrigerated foods. The frozen foods are kept at or below −18 °C prior to final delivery to the consumers (EUR-Lex, 1991). The continuous low temperature environments during the storage and transportation of cold-storage foods create a favorable condition which can drastically prolong the survival of SARS-CoV-2, should the food be contaminated during earlier harvesting or processing (Chin et al., 2020).

There has accumulated growing evidence on SARS-CoV-2 positivity testing from cold-chain food samples or in the packages. But less attention has been paid to frozen and refrigerated foods as a risk factor in COVID-19 risk mitigation strategies (Sun et al., 2021). The high stability of the SARS-CoV-2 virus in refrigeration (below 4 °C) and as frozen (at −10 to −80 °C) has been well established presenting a significant risk to the cold-chain food (WHO, 2020; Han et al., 2021).

In laboratory studies, the risk of cold-chain food for contaminating and spreading of SARS-CoV-2 was investigated. It was observed that the cold-chain food could promote COVID-19 contamination because SARS-CoV-2 remained highly stable under refrigeration at 4 °C, and also at freezing conditions from −10 to −80 °C on fish, meat, poultry, and swine skin, during 14–21 days testing periods (Han et al., 2021, Harbourt et al., 2020, Pang et al., 2020). COVID-19 infectivity on food surfaces has also been reported (Pang et al., 2020). SARS-CoV-2 also remained largely stable on swine skin throughout the 14-day experiment at refrigerating condition of 4 °C (Sun et al., 2021). However, not much data is available concerning the long-term survival and infectivity of the virus under these conditions. Data from the MERS and SARS-CoV-1 epidemics show a potential and continual risk of infections under similar conditions. Bovine CoV-88 retained infectivity for at least 14 days on fresh lettuce in a refrigerator at 4 °C while human coronavirus 229E survived well on lettuce, with just a minimal reduction after 2 days of refrigeration at 4 °C (Han et al., 2021). Therefore, it is important to be aware of the potential risk of spreading the SARS-CoV-2 through contaminated refrigerated or frozen food through the cold-chain.

With growing number of COVID-19 cases in food industry, without appropriate personal protection equipment, non-symptomatic carriers may act as silent vectors and cause clustered infections in enclosed environments (Huff and Singh, 2020; Zuber and Brüssow, 2020).

7.4.4 *Listeria monocytogenes* infestation in fish and meat in Finland

Listeria monocytogenes is a rod-shaped, gram-positive, catalase positive, facultative anaerobic, non-sporulating, psychotrophic mesophilic pathogen. It is responsible for numerous potentially fatal food outbreaks in humans (Freitag et al., 2009, Wilks et al., 2006, Hakalehto, 2015b). Human listeriosis is a relatively rare but serious foodborne disease with a 2018 EU incident rate of 0.47 cases per the population of 100,000 (ECDC, 2021). *L. monocytogenes* is responsible for foodborne listeriosis in humans responsing for about 1% of foodborne illness with high mortality rates. *L. monocytogenes* also continuously presents a growing threat in clinical control due to its growing multidrug resistance in isolates continuously reported in contaminated food products (Hakalehto, 2015a, b Gómez et al., 2014, Hell et al., 2015, Adusei-Mensah et al., 2021). The organism is rather regularly becoming resistant to common antibiotics like oxacillin, clindamycin, and tetracycline (Gómez et al., 2014). It also produces a toxin, listeriocin, which is harming human cells and tissues, as well as the growth and development of the overall microbiome (Rolhion and Cossart, 2017).

L. monocytogenes can also be found in the intestinal tract of about 2–10% of the general population without any apparent health consequences. It can occur intracellularly in human cells, which makes it difficult to get detected. Hence sudden outbreaks of *L. monocytogenes* may implicate increased food contamination, imbalanced gut microbiota, or weakened immune system. The sudden increase in the *Listeria* sp. infections in Finland during the COVID-19 pandemics could be related to the lowered immune protection caused by the virus epidemics. This is more important especially during and beyond the coronavirus pandemics as the virus is believed to have the ability to populate the immune macrophage cells and potentially alter the gut microbiota (Yip et al., 2014). Such ability has been linked to antibody dependent enhancement (ADE) reaction or cytokine storm and to the related worsening situation of the COVID-19 pathogenesis (Yip et al., 2016).

7.4.5 *Listeria* in Finland during the pandemics

As nearly all respiratory tract infections (both viral and bacterial) have shown a significant reduction during the coronavirus pandemic period, *Listeria* infections have almost doubled during the period (Hannila-Handelberg, 2021). The cases are also on higher level than during 1995–2004 with annual number of infections varying between 18 and 53 as average annual cases per 7 million inhabitants (Lyytikäinen et al., 2006). Some common food sources for Listeria are the fresh, unripened cheese and refrigerated fish (Figure 7.7).

Figure 7.7: In the prevention of listeriosis, some fresh foods such as fish need to be carefully stored. In fact, in some studies, 55% of the unripened cheeses in Finland were showed to contain listeria but mostly on very low levels. Photo: Jukka Sauramäki.

According to the Finnish Institute for Health and Welfare's report, a record high of 93 intestinal *Listeria* infections were diagnosed in 2020, which is almost twice as many cases as in 2019 (Hannila-Handelberg, 2021). From the report, dry-cured and cold smoked fish products are the foods at a particular high risk of listerial infections in the country. In addition, gastrointestinal *Campylobacter* infections of domestic origin were also observed to be on elevated levels when compared to the previous years (Hannila-Handelberg, 2021). According to Adj. Prof., Ruska Rimhanen-Finne, "What is remarkable in this cluster is that several [cases of] infections have appeared in a short time in different parts of Finland," *Yle Finnish TV News*, (2020). *Campylobacter* gastrointestinal infections are endemic in Finland and in the year 2020 the total of 2,074 *Campylobacter* infections were reported to the National Infectious Diseases Register (Hannila-Handelberg, 2021).

7.5 Urban hygienic maintenance of the past – or the lack of it

If we position the food deliveries, catering services, and restaurants into the context with time and place, we could steer the "time machine" to London, England, of the eighteenth century and afterwards (White, 2010). The streets were narrow, and what is today's hectic city bank district used to be the cheapside slum area. This neighborhood contained a wide variety of human activities, including numerous public houses, where the food was prepared, distributed, and consumed.

The inhabitants of rapidly growing cities some centuries ago lived in the densely populated areas of Europe having hardly existing sewage systems. In fact, the streets formed the exhaust route for the urban waste, food waste and constituted the wastewater treatment drainage lines. There the occasional rain waters flushed the garbage, cleaning water, human and animal secretions, and other untreated remnants of the hygienically reckless lifestyle into the ditches and finally to such waterways as the river Thames, in the case of the English capital. In a way, this lack of hygienic maintenance was the downside of the rapid urbanization, industrialization, and population growth.

Situation and the circumstances in a pre-modern, industrial revolution time living environment, housing and municipal services, still prevail in numerous cities around the globe. Many of them belong to the so-called megacities, which already in 2013 numbered 29 globally. A megacity refers to a metropole with 10 million or more inhabitants (Johnson, 2019). The population in those areas has to often suffer from almost unbearable air pollution, resembling the famous smog of London (Bell and Davis, 2001). Parallel modern example is the Indian capital New Delhi. There the human population lives in close contact with various animals which increase the potential risks for contagious zoonotic diseases spreading or developing into pandemics.

The air pollution is causing an incessant respiratory exposure with smoke, toxic fumes, dust, aerosols, and microorganisms. This undoubtedly increases the risk of epidemics. "Seasonal" flu outbreaks seem to have been common throughout the second half of the eighteenth century London (White, 2010). The influenza of 1762 "pervaded the whole city far and wide, scarcely sparing anyone."

The corresponding devastating Spanish flu (caused by H1N1 influenza virus), followed by pneumococcal and other bacterial infections in the Winter of 1918–19 killed as many as 4,000 people each week in London. Already in 1710 in the much smaller town of Helsinki, 1,185 city dwellers were buried into the old cemetery within four months. Then the horrible plague killed two thirds of the local population (Frandsen, 2009).

7.6 Parallel situation in today's world – pandemics abruptly influence the societies

In human lungs, the narrow airways caused obstruction when viral or bacterial infection is initiating the swelling of the tissues (Jaakkola and Hakalehto, 2015). This boosts the worsening of the respiratory diseases which outcome is further provoked by dirty polluted air and poor hygienic conditions. The infection and inflammation then lower the lung ventilation rate which initiates a vicious circle. Many pathogens are potentially sneaking into the lungs and could initiate an infection as the overall and respiratory health of an individual is worsening during the pandemics.

In fact, tuberculosis is a lung disease caused by a bacterium, *Mycobacterium tuberculosis*, which is estimated to have caused at least as many deaths in human history as the other contagious diseases together (Hakalehto, 2015c; 2015d). This infective agent is widely spread all over, and every fourth citizen of our global village is a carrier of this microaerobic germ. Tuberculosis (TB) and other bacterial complications are likely consequences of the COVID-19 virus pandemics and other viral epidemics. Fortunately, it is possible to monitor this bacterium and other slowly or fast propagating pathogens with the portable microbe enrichment unit (PMEU) device (Hakalehto, 2013; Hakalehto et al., 2014). Moreover, the antibiotic resistant variants should be detected.

Pandemic situation changed practically all transportation operations in many parts of the world; sharply grown ecommerce, global trade, closures of businesses and societies, and it increased the need for special transportations. Also, the topic that influenced all services during the exceptional situation is the urgent need for protection of workers, disinfection of hands, equipment, and goods and special orders how to avoid all contacts and keep the required distancing.

Pandemics also dramatically changed the distribution of the food, the catering services, and restaurant home deliveries. Remote work recommendations and national or aerial lockdowns forced restaurants to launch home delivery services and protection of elderly people set new standards for social catering services to individuals living at their homes and to institutions like nursing homes and other service units.

Delivery and last mile drivers face the recipient usually dozens of times during the day. When the pandemic was at its worst state, these situations and encounters caused a clear risk to the driver as well as to the recipient. Daily routes in operational transportation units in distribution centers create the network of possible infection points, and terminals and hubs where drivers work together and with indoor staff and management, enable the spread of infection within the work community.

The risks can be easily described also to have emerged to another direction. If the driver is infected, they are the potential super-spreaders contacting number of recipients, visiting working places, and terminals and in the worst-case nursing homes where the residents are in the most vulnerable position. These risks exist at both directions, and drivers' role as mobile service person clearly position food delivery personnel, among other last mile operators, to the frontline in encountering the pandemic situation.

To avoid above-mentioned risks, distribution operators have been forced to implement various methods to protect their personnel as well as to participate in societal protection projects. Now, these methods and equipment sound the most obvious things to be implemented but before COVID-19 pandemic era they were often considered as over-reaction, too difficult to be implemented or executed, and at least mitigating the overall efficiency of the services. However, prudent hygienic practices could make the business flourish, as the customers recognize the service to reach high levels.

7.7 PMEU – portable microbe enrichment unit

The PMEU device (Finnoflag Oy, Kuopio, Finland) (Hakalehto, 2009; Hakalehto and Heitto, 2012) is a portable incubator where the enrichment of microbes takes place in syringes containing enrichment broth. Gas flow is funneled into syringes through a sterile filter and a needle to agitate the broth and keep the bacterial cells in movement. Different gases or gas mixtures are used for aerobic and anaerobic enrichment.

The PMEU device has been designed to promote recovery and accelerate the growth of microbes under aseptic gas flow in adjusted temperatures (Pitkänen et al., 2009). It reveals the enteric and other bacteria in clinical, industrial, and environmental samples and has also been used for the investigations on the properties of enteric bacteria and interactions between members of the intestinal flora (Hakalehto, 2006, 2011; Hakalehto et al., 2007–2009; Pesola et al., 2009). The PMEU method has earlier been validated for the detection of coliformic bacteria (Wirtanen and Salo, 2010) and for the enrichment and detection of *Campylobacter* sp. in waters (Pitkänen et al., 2009). For more information on the PMEU technology, see Chapter 6.

7.8 There are many variations of the theme – the preventive and protective measures should be variable and flexible enough

Governments and health officials instructed citizens and businesses during the first months of pandemic how to protect and get protection. These instructions were implemented soon in the delivery operations globally, and growing general awareness of risks and threats also made the customers to demand sustainability from the logistics industry also in this way. Occasional shortages of protective gears and hand sanitizers caused an increase in operational costs, but soon these protection methods became a new normal. The basic toolbox of driver's protection includes face masks, gloves, and hand sanitizer, more or less the same equipment as recommended for private citizens by health officials in many countries.

More complicated task has been to find ways to serve customers and distribute food and at same time avoid unnecessary contacts. In Finland, municipalities encourage elderly people to continue living at their homes, if possible, by offering services like food delivery. Delivering their daily portions, the driver also checks out that everything is fine with the customer and reports home care unit if some actions are needed. This is the task that must be executed, but it also needs to be performed carefully and discreetly with well protected personnel. In many cases, the driver is one of the rare persons that the customer meets daily, if not the only one. This contact is important, and it was even more important during times of isolation that were declared to protect senior citizens in many countries.

Food deliveries to hospitals, nursing homes, and other institutions have also demanded new operational instructions and guidelines. Before the pandemic situation, there weren't strict rules existing how to hand over food distribution carts and boxes. It was more a logistics issue how far from the loading dock transportation units were taken by the driver. Awareness on COVID-19 risks changed these instructions. To protect patients and residents of these institutions, drivers were instructed to leave incoming units outside the care departments. To protect drivers from possible infections from the hospitals, extra caution is needed. Outgoing units were left in a separate place for disinfection before loading the vehicle.

7.9 Onset of the pandemics and subsequent measures

The start of the pandemic was a huge surprise to almost every citizen in the world. Some health professionals, microbiologists, and virologists warned the world during and after previous threatening diseases like bird flu and SARS-CoV-1, but

Figure 7.8: Home deliveries and all customers contacts have been made with extra attention to safety.

practically all logistics companies in the Western world were unprepared for this kind of situation. In Finland, the leading postal and logistics service provider Posti Group Oyj is also the major operator in food deliveries. In very early stages of COVID-19, the management of Posti formed a cross-functional team to make immediate decisions how to prepare and operate the operational functions in exceptional circumstances (Figure 7.8). Team included all operational units, all business units, well-being, and work safety specialists, and it was supported by occupational health professionals from private healthcare service providers.

Posti Ltd. has more than 10,000 operational workers, and there are additionally thousands more who work for the company by sub-contractors. There are also hundreds of supervisors, transport operators, field managers, and others in supporting roles. There was urgent need to publish updated work instructions, safety rules, and even guidelines for safe commuting. All these instructions were made according to the guidelines of the national health authorities. When there were changes in these guidelines, there appeared a need for new instructions on the company level, too. It was a very complicated issue to instruct big operational units in the circumstances that were strange to all. More specialists and relevant knowledge providers should be involved in strategic decision-making. In case of the dangerous diseases threatening the society, more microbiologists and other specialists would be urgently needed to join the operational teams.

Sustainability in business language used to mean company's efforts to reduce emissions, battling climate change and supporting equality among workers, valuing diversity etc. The pandemic made companies to think one more subject. How do we protect our workers and how can we protect our customers and even societies as a whole? Posti Ltd, as a highly ranked sustainable company (https://www.posti.com/en/sustainability/news/ecovadis/), has been searching for new methods and technologies to take care of the safety of her employees and customers. One result of this search is the implementation of the UVC light disinfection technology in disinfecting mobile phones and devices. Some of these devices are used by many workers during the shift and some of them will normally be handled by customers, too. By using UVC light, it is possible to disinfect devices quickly, effectively and without damaging the sensitive technology of the equipment.

The pandemic wave has been disastrous in many parts of the world and for many individuals in the countries that have suffered less in general. Still, for service providers, like many logistics and fulfillment companies, it has taught a valuable lesson. Paying an attention to microbiological safety of workers, they can do a lot for the well-being of their employees, reduce the cost of sick leaves, and minimize the quality issues caused by the absence of top professionals.

7.10 Conclusions

The COVID-19 global crisis was, and still is, a sudden alert for the authorities and businesses in all countries. The food distribution sector was one of the areas which were the first ones to encounter the partially unknown pandemic threat. This situation underlined the need for better preparedness for sudden outbreaks of diseases in future.

Moreover, the hygiene maintenance operations during the food chains need to be continuously re-evaluated. In exceptional times, the security of supplies is on the shoulders of few. Therefore, careful precautions are warranted, and should be designed during more stable periods of time.

References

Adusei-Mensah, F., Hakalehto, E., Tikkanen-Kaukanen, C. (2021). Microbiological and Chemical Safety of African Herbal and Natural Products. Germany: De Gruyter Publishers.
Armon, R. (2015). Nosocomial viruses. In: Hakalehto, E. (editor). Microbiological Clinical Hygiene. New York, NY, USA: Nova Science Publishers, Inc., 35–51.
Bahl, P., Doolan, C., De Silva, C., Chughtai, A., Bourouiba, L., MacIntyre, C. (2020). Airborne or droplet precautions for health workers treating COVID-19?. Journal of Infectious Disease Apr; 16: 189. doi: 10.1093/infdis/jiaa189.

Bell, M. L., Davis, D. L. (2001). Reassessment of the lethal London fog of 1952: Novel indicators of acute and chronic consequences of acute exposure to air pollution. Environmental Health Perspectives 2001 Jun; 109 (Suppl 3): 389–394. doi: 10.1289/ehp.01109s3389.

Casella, M. L., Schmidt-Lorenz, W. (1989). Disinfection with gaseous formaldehyde. First Part: Bactericidal and sporicidal effectiveness of formaldehyde with and without formation of a condensing layer. Zentralbl Hyg Umweltmed May; 188 (1–2): 144–165.

Chin, A. W. H., Chu, J. T. S., Perera, M. R. A., Hui, K. P. Y., Yen, H.-L., Chan, M. C. W., Peiris, M., Poon, L. L. M. (2020). Stability of SARS-CoV-2 in different environmental conditions. MedRxiv, 2020.03.15. 20036673. doi: https://doi.org/10.1101/2020.03.15.20036673.

ECDC. (2021). Seventh external quality assessment scheme for Listeria monocytogenes typing. European Centre for Disease Prevention and Control. https://www.ecdc.europa.eu/en/publications-data/listeria-monocytogenes-typing-seventh-external-quality-assessment-scheme

EUR-Lex. (1991). Council Directive 89/108/EEC of 21 December 1988 on the approximation of the laws of the Member States relating to quick-frozen foodstuffs for human consumption. https://eur-lex.europa.eu/legal-content/en/ALL/?uri=CELEX%3A31989L0108

Freitag, N. E., Port, G. C., Miner, M. D. (2009). Listeria monocytogenes – From saprophyte to intracellular pathogen. Nature Reviews Microbiology, 7 (9): 623–628. doi: https://doi.org/10.1038/nrmicro2171.

Frandsen, K.-E. (2009). The Last Plague in the Baltic Region. 1709–1713. Copenhagen. ISBN 9788763507707.

Gómez, D., Azón, E., Marco, N., Carramiñana, J. J., Rota, C., Ariño, A., Yangüela, J. (2014). Antimicrobial resistance of Listeria monocytogenes and Listeria innocua from meat products and meat-processing environment. Food Microbiology, 42: 61–65. doi: https://doi.org/10.1016/j.fm.2014.02.017.

Gundy, P. M., Gerba, C. P., Pepper, I. L. (2008). Survival of coronaviruses in water and wastewater. Food and Environmental Virology, 1 (1): 10. doi: https://doi.org/10.1007/s12560-008-9001-6.

Hakalehto, E. (2006). Semmelweis' present day follow-up: Updating bacterial sampling and enrichment in clinical hygiene. Pathophysiology, 13: 257–267. doi: http://dx.doi.org/10.1016/j.pathophys.2006.08.004.

Hakalehto, E. (2011). Simulation of enhanced growth and metabolism of intestinal Escherichia coli in the Portable Microbe Enrichment Unit (PMEU). In: Rogers, M. C., Peterson, N. D. (eds.). E. Coli Infections: Causes, Treatment and Prevention. New York, USA: Nova Science Publishers, 159–175.

Hakalehto, E. (2013). Interactions of Klebsiella sp. with other intestinal flora. In: Pereira, L. A., Santos, A. (editors). Klebsiella Infections: Epidemiology, Pathogenesis and Clinical Outcomes. New York, USA: Nova Science Publishers, Inc., 2013, 1–33.

Hakalehto, E. (2015–2018). Editor in Forthcoming series: 1. Microbiological clinical hygiene, 2. Microbiological food hygiene, 3. Microbiological industrial hygiene, 4. Microbiological Environmental Hygiene.

Hakalehto, E. (2015a). Antibiotic resistance in foods. In: Hakalehto, E. (ed.). Microbiological Food Hygiene, 1st edn, ed., p. 43. New York: Nova Publishers, 233–235.

Hakalehto, E. (2015b). Hazards and prevention of food spoilage. In: Hakalehto, E. (ed.). Microbiological Food Hygiene, 1st edn, ed., p. 43. New York: Nova Publishers.

Hakalehto, E. (2015c). Mycobacterial detection views. In: Hakalehto, E. (ed). Microbiological Clinical Hygiene. New York, NY, USA: Nova Science Publishers, Inc, 135–143.

Hakalehto, E. (2015d). Hygienic lessons from the dairy microbiology cases. In: Hakalehto, E. (ed.). Microbiological Food Hygiene. New York, NY, USA: Nova Science Publishers, Inc, 155–174.

Hakalehto, E. (2020). Current megatrends in food production related to microbes. Journal of Food Chemistry and Nanotechnology, 6 (1): 78–87.

Hakalehto, E., Väätäinen, U. (2016). Hygienic aspects in healthcare industries and services. In: Hakalehto, E. (ed.). Microbiological Industrial Hygiene. New York, NY, USA: Nova Science Publishers, Inc.

Hakalehto, E., Humppi, T. (2018). Monitoring microbes in the ambient air. In: Hakalehto, E. (ed.). Microbiological Environmental Hygiene. New York, NY, USA: Nova Science Publishers, Inc., 383–404.

Hakalehto, E., Heitto, A., Heitto, L., Rissanen, K., Pesola, I., Pesola, J. (2014). Enhanced recovery, enrichment and detection of Mycobacterium marinum with the Portable Microbe Enrichment Unit (PMEU). Pathophysiology, 21: 231–235. doi: 10.1016/j.pathophys.2014.07.005. Epub 2014 Aug 4.

Hakalehto, E., Heitto, L. (2012). Minute microbial levels detection in water samples by Portable Microbe Enrichment Unit Technology. Environment and Natural Resources Research, 2: 80–88.

Hakalehto, E., Hell, M., Heitto, A. (2015). Planning and construction of healthcare facilities – a case evaluation on hygienic aspects. In: Hakalehto, E. (editor). Microbiological Clinical Hygiene. New York, NY, USA: Nova Science Publishers, Inc., 79–84.

Hakalehto, E., Humppi, T., Paakkanen, H. (2008). Dualistic acidic and neutral glucose fermentation balance in small intestine: Simulation in vitro. Pathophysiology, 15: 211–220. doi: http://dx.doi.org/10.1016/j.pathophys.2008.07.001.

Hakalehto, E., Pesola, J., Heitto, A., Bhanj Deo, B., Rissanen, K., Sankilampi, U., Humppi, T., Paakkanen, H. (2009). Fast detection of bacterial growth by using Portable Microbe Enrichment Unit (PMEU) and ChemPro100i® gas sensor. Pathophysiology, 16: 57–62. doi: http://dx.doi.org/10.1016/j.pathophys.2009.03.001.

Hakalehto, E., Pesola, J., Heitto, L., Narvanen, A., Heitto, A. (2007). Aerobic and anaerobic growth modes and expression of type 1 fimbriae in Salmonella. Pathophysiology, 14: 61–69. doi: http://dx.doi.org/10.1016/j.pathophys.2007.01.003.

Hallak, J., Teixeira, T.A., Bernardes, F.S., et al. (2021). SARS-CoV-2 and its relationship with the genitourinary tract: Implications for male reproductive health in the context of COVID-19 pandemic. Andrology. 9: 73–79. https://doi.org/10.1111/andr.12896.

Han, J., Zhang, X., He, S., Jia, P. (2021). Can the coronavirus disease be transmitted from food? A review of evidence, risks, policies and knowledge gaps. Environmental Chemistry Letters, 19 (1): 5–16. doi: https://doi.org/10.1007/s10311-020-01101-x.

Hannila-Handelberg, T. (2021). COVID-19 restrictions also notably reduced other respiratory tract infections in 2020 – But listeria infections at record high – Press release – THL. Finnish Institute for Health and Welfare (THL), Finland. https://thl.fi/en/web/thlfi-en/-/COVID-19-restrictions-also-notably-reduced-other-respiratory-tract-infections-in-2020-but-listeria-infections-at-record-high

Harbourt, D. E., Haddow, A. D., Piper, A. E., Bloomfield, H., Kearney, B. J., Fetterer, D., Gibson, K., Minogue, T. (2020). Modeling the stability of severe acute respiratory syndrome coronavirus 2 (SARS-CoV-2) on skin, currency, and clothing. MedRxiv, 2020. 07.01.20144253. doi: https://doi.org/10.1101/2020.07.01.20144253.

Hell, M., Bernhofer, C., Pesola, J., Pesola, I., Hakalehto, E. (2015). Prevalence, detection and prevention of foodborne outbreaks related to large hospital kitchens. In: Hakalehto, E. (ed.). Microbiological Food Hygiene, 1st edn. New York: Nova Publishers, 96.

Hokajärvi, A.-M., Rytkönen, A., Tiwari, A., Kauppinen, A. (2021). The detection and stability of the SARS-CoV-2 RNA biomarkers in wastewater influent in Helsinki, Finland. Science of the Total Environment. doi: 10.1016/j.scitotenv.2021.145274.

https://thl.fi/en/web/infectious-diseases-and-vaccinations/what-s-new/coronavirus-COVID-19-latest-updates/situation-update-on-coronavirus

https://www.posti.com/en/sustainability/news/ecovadis/

Huang, C., Huang, L., Wang, Y., Li, X., Ren, L., Gu, X., et al. (2021). 6-month consequences of COVID-19 in patients discharged from hospital cohort study. Lancet Jan 08; 397 (Issue 10270): p220–232.o.

Huff, V., Singh, A. (2020). Asymptomatic transmission during the coronavirus disease. Pandemic and implications for public health strategies. Clinical Infectious Disease, 71 (10): 2752–2756. doi: https://doi.org/10.1093/cid/ciaa654.

Immonen, M., Hakalehto, J.-P., Hakalehto, E. (2015). Trends toward clean and healthy nutrition. In: Hakalehto, E. (ed.). Microbiological Food Hygiene. New York, NY, USA: Nova Science Publishers, Inc, 19–31.

Jaakkola, K., Hakalehto, E. (2015). Some common mild and more serious respiratory infections. In: Hakalehto, E. (ed.). Microbiological Clinical Hygiene. New York, NY, USA: Nova Science Publishers, Inc, 53–77.

Johnson, H. (2019). These destinations will earn megacity status by 2030. CEO Magazine.

Karl-Erik, F. (2009). The Last Plague in the Baltic Region. 1709–1713, 20. Copenhagen: Museum Tusculanum Press.

Kreier, F. (2021). The myriad ways sewage surveillance is helping fight COVID around the world. Nature. doi: https://doi.org/10.1038/d41586-021-01234-1.

La Rosa, G., Bonadonna, L., Lucentini, L., Kenmoe, S., Suffredini, E. (2020). Coronavirus in water environments: Occurrence, persistence and concentration methods – A scoping review. Water Research, 179: 115899. doi: https://doi.org/10.1016/j.watres.2020.115899.

Lineback, C. B., Nkemngong, C. A., Wu, S. T., Li, X., Teska, P. J., Oliver, H. F. (2018). Hydrogen peroxide and sodium hypochlorite disinfectants are more effective against Staphylococcus aureus and Pseudomonas aeruginosa biofilms than quaternary ammonium compounds. Antimicrob Resistance and Infection Control Dec; 17 (7): 154. doi: 10.1186/s13756-018-0447-5.

Lyytikäinen, O., Nakari, U. M., Lukinmaa, S., Kela, E., Minh, N. N. T., Siitonen, A. (2006). Surveillance of listeriosis in Finland during 1995–2004. Eurosurveillance, 11 (6): 5–6. doi: https://doi.org/10.2807/esm.11.06.00630-en.

Mahama, J. (2020). COVID-19 is a chance for Africa to have its own Marshall plan – but not as we know it | opinion. Newsweek, 26 (5): 2020.

Majumder, J., Minko, T. (2021). Recent developments on therapeutic and diagnostic approaches for COVID-19. AAPS Journal 2021 Jan 5; 23 (1): 14. doi: 10.1208/s12248-020-00532-2.

Morens, D. M., Taubenberger, J. K., Fauci, A. S. (2008). Predominant role of bacterial pneumonia as a cause of death in pandemic influenza: Implications for pandemic influenza preparedness. Journal of Infectious Disease Oct 1; 198 (7): 962–970. doi: 10.1086/591708.

Östergaard, L. (2021). SARS CoV-2 related microvascular damage and symptoms during and after COVID-19: Consequences of capillary transit-time changes, tissue hypoxia and inflammation. Physiological Reports, 9: e14726.

Pang, X., Ren, L., Wu, S., Ma, W., Yang, J., Di, L., Li, J., Xiao, Y., Kang, L., Du, S., Du, J., Wang, J., Li, G., Zhai, S., Chen, L., Zhou, W., Lai, S., Gao, L., Pan, Y., Wang, Q., Li, M., Wang, J., Huang, Y., Wang, J. (2020). COVID-19 Laboratory Testing Group. (2020). Cold-chain food contamination as the possible origin of COVID-19 resurgence in Beijing. National Science Review, 7 (12): 1861–1864. doi: https://doi.org/10.1093/nsr/nwaa264.

Pesola, J., Vaarala, O., Heitto, A., Hakalehto, E. (2009). Use of portable enrichment unit in rapid characterization of infantile intestinal enterobacterial microbiota. Microbial Ecology in Health and Disease, 21: 203–210. doi: http://dx.doi.org/10.3109/08910600903367810.

Pitkänen, T. (2020). Coronavirus detected in the wastewater of several cities in August – Press release – Finnish Institute for Health and Welfare (THL), Finland. https://thl.fi/en/web/thlfi-en/-/coronavirus-detected-in-the-wastewater-of-several-cities-in-august

Pitkänen, T., Bräcker, J., Miettinen, I. T., Heitto, A., Pesola, J., Hakalehto, E. (2009). Enhanced enrichment and detection of thermotolerant Campylobacter species from water using the Portable Microbe Enrichment Unit and real-time PCR. Canadian Journal of Microbiology, 55 (7): 849–858.

Richardson, H. (2021). How wastewater is helping South Africa fight COVID-19. Nature, 593 (7860): 616–617. doi: https://doi.org/10.1038/d41586-021-01399-9.

Rolhion, N., Cossart, P. (2017). How the study of Listeria monocytogenes has led to new concepts in biology. Future Microbiology Jun; 12: 621–638. doi: 10.2217/fmb-2016-0221. Epub 2017 Jun 12.

Sauramäki, J., Hakalehto, E. (2015). Catering services and hygienic food deliveries. Posti Ltd, Helsinki, Finland, and others. In: Hakalehto, E. (ed.). Microbiological Environmental Hygiene. Chapter 17. New York, NY, USA: Nova Science Publishers, Inc.

Sun, C., Cheng, C., Zhao, T., Chen, Y., Ayaz Ahmed, M. (2021). Frozen food: Is it safe to eat during COVID-19 pandemic?. Public Health, 190: e26. doi: https://doi.org/10.1016/j.puhe.2020.11.019.

Tian, Y., Zhou, L. Q. (2021). Evaluating the impact of COVID-19 on male reproduction. Reproduction Feb; 161 (2): R37–R44. doi: 10.1530/REP-20-0523.

Wang, X.-W., Li, J.-S., Jin, M., Zhen, B., Kong, Q.-X., Song, N., Xiao, W.-J., Yin, J., Wei, W., Wang, G.-J., Si, B., Guo, B.-Z., Liu, C., Ou, G.-R., Wang, M.-N., Fang, T.-Y., Chao, F.-H., Li, J.-W. (2005). Study on the resistance of severe acute respiratory syndrome-associated coronavirus. Journal of Virological Methods, 126 (1–2): 171–177. doi: https://doi.org/10.1016/j.jviromet.2005.02.005.

White, J. (2010). London – The Story of a Great City. Andre Deutsch, London, U.K.

Wilks, S. A., Michels, H. T., Keevil, C. W. (2006). Survival of Listeria monocytogenes Scott A on metal surfaces: Implications for cross-contamination. International Journal of Food Microbiology, 111 (2): 93–98. doi: https://doi.org/10.1016/j.ijfoodmicro.2006.04.037.

Wirtanen, G., Salo, S. (2010). PMEU-laitteen validointi ulostebakteereilla (in Finnish, Validation of PMEU device with fecal bacteria). Espoo, Finland: VTT Expert Services Oy.

World Health Organizaton (2020). Coronavirus disease 2019 (COVID-19) Situation Report – 32. HIGHLIGHTS. https://www.who.int/docs/default-source/coronaviruse/situation-reports/20200221-sitrep-32-COVID-19.pdf?sfvrsn=4802d089_2

Yip, M. S., Leung, H. L., Li, P. H., Cheung, C. Y., Dutry, I., Li, D., Daëron, M., Bruzzone, R., Peiris, J. S., Jaume, M. (2016). Antibody-dependent enhancement of SARS coronavirus infection and its role in the pathogenesis of SARS. Hong Kong Medical Journal = Xianggang Yi Xue Za Zhi, 22 (3 Suppl 4): 25–31.

Yip, M. S., Leung, N. H. L., Cheung, C. Y., Li, P. H., Lee, H. H. Y., Daëron, M., Peiris, J. S. M., Bruzzone, R., Jaume, M. (2014). Antibody-dependent infection of human macrophages by severe acute respiratory syndrome coronavirus. Virology Journal, 11: 82. doi: https://doi.org/10.1186/1743-422X-11-82.

Yle News (2021). THL probing listeria infections in Finland. (2020). TV News. https://yle.fi/uutiset/osasto/news/thl_probing_listeria_infections_in_finland/11453495

Yuan, J., Chen, Z., Gong, C., Liu, H., Li, B., Li, K., Chen, X., Xu, C., Jing, Q., Liu, G., Qin, P., Liu, Y., Zhong, Y., Huang, L., Zhu, B.-P., Yang, Z. (2020). Sewage as a possible transmission vehicle during a coronavirus disease 2019 outbreak in a densely populated community: Guangzhou, China, April 2020. Clinical Infectious Diseases, ciaa1494. doi: https://doi.org/10.1093/cid/ciaa1494.

Zuber, S., Brüssow, H. (2020). COVID 19: Challenges for virologists in the food industry. Microbial Biotechnology, 13 (6): 1689–1701. doi: https://doi.org/10.1111/1751-7915.13638.

About the editor

Dr. E. Elias Hakalehto, Ph.D., graduated as an environmental microbiologist from the University of Helsinki, Finland in 1983. During his student years, he acted as a Research Trainee at the Biotechnology Laboratory of the VTT State Research Centre of Finland. In 1983–84, he was elected to the Board of the Student Union of Helsinki University, where he also established the Environmental Section, being its first chairman. He completed the postgraduate diploma in biotechnology at the University of Kent at Canterbury, UK, in 1985. He is also an alumnus of the University College London (in biochemical engineering). At Westminster University (then Polytechnic of Central London), he prepared a research work on "Gene cloning in *Streptomyces*." This lab work constructed the first chimeric vector for gene transfer between antibiotics producing *Streptomycetes* and *Escherichia coli*. In 1985, he was the secretary of the global MIRCEN congress of the Microbial Research Centers in Helsinki. In 1986, he was the first researcher in a joint project of the University of Helsinki and Valio Dairies Corporation, focusing on implementing probiotics into the company's product portfolio. As a result, Valio purchased the GG strain (DSM 33156) and made it a leading concept in the field under several trade names, such as Gefilus™ in Finland and Culturelle™ in the USA. In the 1980s, Elias Hakalehto served as a teacher and course leader in the Universities of Helsinki and Eastern Finland (then Kuopio University), running and designing courses in biotechnology, food and environmental microbiology, etc. He acted as a visiting scientist in the University of Jerusalem (Biotechnology Unit and Hadassah Medical School) in 1989–90. During 1992–95, he served as a project leader in the projects "Rapid Microbial Detection," funded by the Finnish National Fund, and "Bioagent Threat Monitoring" for the Finnish Defence Forces. His academic dissertation (Ph.D.) was presented to the University of Kuopio (now University of Eastern Finland). Dr. Hakalehto has lectured in numerous courses and conferences. For example, he has given lectures to the students of Savonia University of Applied Sciences for five decades. During 1997–2002, Dr. Hakalehto was a member and secretary of the Toxicology and Microbiology Section of the Finnish Committee of National Defense. To this day, his R&D company, Finnoflag Oy (est.1993), has conducted more than 100 research projects for academic and industrial clients, including more than 10 semi-industrial pilot studies. His invention, the PMEU (Portable Microbe Enrichment Unit), was the key technique in the nationwide Polaris project with 29 participating institutions (2009–2012) for automatically monitoring the microbiological hygiene levels of the water distribution networks. Finnoflag Oy and Dr. Hakalehto were acting as key technology providers also in the European Union Baltic Sea region six-nation biorefinery pilot project ABOWE in 2012–14 (Germany, Poland, Sweden, Finland, Lithuania, and Estonia), and in the "Zero Waste from Zero Fiber" project ordered by the City of Tampere and funded by the Finnish Ministry of Agriculture and Forestry in 2018–19. Dr. Elias Hakalehto was appointed as the Adjunct Professor at the University of Kuopio (now University of Eastern Finland) in biotechnical microbe analytics (2008–), and at the University of Helsinki in microbiological agroecology (2016–). He has more than 60 peer-reviewed publications, over 70 original patent applications, and about 120 chapters in scientific books. He has edited more than five scientific books in the fields of microbiology and biotechnology. Since 2015, Dr. Hakalehto has served as the Vice President of the International Society of Environmental Indicators (ISEI), acting as the chairman of the 22nd global ICEI conference in Helsinki in 2017. He has been the keynote speaker in numerous conferences and webinars, such as FCT (Food Chemical Technology) in Baltimore (in 2017) and Los Angeles (in 2019), USA. Starting from San Francisco in 2018, he has given, as the CEO of his company Finnoflag Oy, keynote lectures also at several conferences on probiotics. He was the Chairman of the Microbiology session in the 2019 International Conference in Microbiology Research and Applications in Baltimore, also organized by the United Scientific Inc. such as the FCT meetings.

https://doi.org/10.1515/9783110724967-009

Dr. Hakalehto chaired in June 2020 the webinar on functional foods and immune systems (WFFIS). He has belonged to the organizing committee of more than 10 international scientific events. In May 2021, he lectured at the Autoimmunity 2021 web conference to a wide global audience. Since 2021 he has been the Fellow Member of the International Society for Development and Sustainability (ISDS) (Japan).

Index

https://doi.org/10.1515/9783110724967-010

www.ingramcontent.com/pod-product-compliance
Lightning Source LLC
Chambersburg PA
CBHW081527220326
41598CB00036B/6359